LONDON MATHEMATICAL SOCIETY LECTURE NOTE SERIES

Managing Editor: Professor J.W.S. Cassels,
Department of Pure Mathematics and Mathematical Statistics,
16 Mill Lane, Cambridge CB2 1SB.

1. General cohomology theory and K-theory, P.HILTON
4. Algebraic topology, J.F.ADAMS
5. Commutative algebra, J.T.KNIGHT
8. Integration and harmonic analysis on compact groups, R.E.EDWARDS
9. Elliptic functions and elliptic curves, P.DU VAL
10. Numerical ranges II, F.F.BONSALL & J.DUNCAN
11. New developments in topology, G.SEGAL (ed.)
12. Symposium on complex analysis, Canterbury, 1973, J.CLUNIE & W.K.HAYMAN (eds.)
13. Combinatorics: Proceedings of the British Combinatorial Conference 1973, T.P.McDONOUGH & V.C.MAVRON (eds.)
15. An introduction to topological groups, P.J.HIGGINS
16. Topics in finite groups, T.M.GAGEN
17. Differential germs and catastrophes, Th.BROCKER & L.LANDER
18. A geometric approach to homology theory, S.BUONCRISTIANO, C.P. ROURKE & B.J.SANDERSON
20. Sheaf theory, B.R.TENNISON
21. Automatic continuity of linear operators, A.M.SINCLAIR
23. Parallelisms of complete designs, P.J.CAMERON
24. The topology of Stiefel manifolds, I.M.JAMES
25. Lie groups and compact groups, J.F.PRICE
26. Transformation groups: Proceedings of the conference in the University of Newcastle-upon-Tyne, August 1976, C.KOSNIOWSKI
27. Skew field constructions, P.M.COHN
28. Brownian motion, Hardy spaces and bounded mean oscillations, K.E.PETERSEN
29. Pontryagin duality and the structure of locally compact Abelian groups, S.A.MORRIS
30. Interaction models, N.L.BIGGS
31. Continuous crossed products and type III von Neumann algebras, A.VAN DAELE
32. Uniform algebras and Jensen measures, T.W.GAMELIN
33. Permutation groups and combinatorial structures, N.L.BIGGS & A.T.WHITE
34. Representation theory of Lie groups, M.F. ATIYAH <u>et al.</u>
35. Trace ideals and their applications, B.SIMON
36. Homological group theory, C.T.C.WALL (ed.)
37. Partially ordered rings and semi-algebraic geometry, G.W.BRUMFIEL
38. Surveys in combinatorics, B.BOLLOBAS (ed.)
39. Affine sets and affine groups, D.G.NORTHCOTT
40. Introduction to Hp spaces, P.J.KOOSIS
41. Theory and applications of Hopf bifurcation, B.D.HASSARD, N.D.KAZARINOFF & Y-H.WAN
42. Topics in the theory of group presentations, D.L.JOHNSON
43. Graphs, codes and designs, P.J.CAMERON & J.H.VAN LINT
44. Z/2-homotopy theory, M.C.CRABB
45. Recursion theory: its generalisations and applications, F.R.DRAKE & S.S.WAINER (eds.)
46. p-adic analysis: a short course on recent work, N.KOBLITZ
47. Coding the Universe, A.BELLER, R.JENSEN & P.WELCH
48. Low-dimensional topology, R.BROWN & T.L.THICKSTUN (eds.)
49. Finite geometries and designs, P.CAMERON, J.W.P.Hirschfield & D.R.Hughes (eds.)

50. Commutator calculus and groups of homotopy classes, H.J.BAUES
51. Synthetic differential geometry, A.KOCK
52. Combinatorics, H.N.V.TEMPERLEY (ed.)
53. Singularity theory, V.I.ARNOLD
54. Markov processes and related problems of analysis, E.B.DYNKIN
55. Ordered permutation groups, A.M.W.GLASS
56. Journées arithmétiques 1980, J.V.ARMITAGE (ed.)
57. Techniques of geometric topology, R.A.FENN
58. Singularities of smooth functions and maps, J.MARTINET
59. Applicable differential geometry, M.CRAMPIN & F.A.E.PIRANI
60. Integrable systems, S.P.NOVIKOV et al.
61. The core model, A.DODD
62. Economics for mathematicians, J.W.S.CASSELS
63. Continuous semigroups in Banach algebras, A.M.SINCLAIR
64. Basic concepts of enriched category theory, G.M.KELLY
65. Several complex variables and complex manifolds I, M.J.FIELD
66. Several complex variables and complex manifolds II, M.J.FIELD
67. Classification problems in ergodic theory, W.PARRY & S.TUNCEL
68. Complex algebraic surfaces, A.BEAUVILLE
69. Representation theory, I.M.GELFAND et al.
70. Stochastic differential equations on manifolds, K.D.ELWORTHY
71. Groups - St Andrews 1981, C.M.CAMPBELL & E.F.ROBERTSON (eds.)
72. Commutative algebra: Durham 1981, R.Y.SHARP (ed.)
73. Riemann surfaces: a view towards several complex variables, A.T.HUCKLEBERRY
74. Symmetric designs: an algebraic approach, E.S.LANDER
75. New geometric splittings of classical knots (algebraic knots), L.SIEBENMANN & F.BONAHON
76. Linear differential operators, H.O.CORDES
77. Isolated singular points on complete intersections, E.J.N.LOOIJENGA
78. A primer on Riemann surfaces, A.F.BEARDON
79. Probability, statistics and analysis, J.F.C.KINGMAN & G.E.H.REUTER (eds.)
80. Introduction to the representation theory of compact and locally compact groups, A.ROBERT
81. Skew fields, P.K.DRAXL
82. Surveys in combinatorics: Invited papers for the ninth British Combinatorial Conference 1983, E.K.LLOYD (ed.)
83. Homogeneous structures on Riemannian manifolds, F.TRICERRI & L.VANHECKE
84. Finite group algebras and their modules, P.LANDROCK
85. Solitons, P.G.DRAZIN
86. Topological topics, I.M.JAMES (ed.)
87. Surveys in set theory, A.R.D.MATHIAS (ed.)
88. FPF ring theory, C.FAITH & S.PAGE
89. An F-space sampler, N.J.KALTON, N.T.PECK & J.W.ROBERTS
90. Polytopes and symmetry, S.A.ROBERTSON
91. Classgroups of group rings, M.J.TAYLOR
92. Representation of rings over skew fields, A.H. SCHOFIELD
93. Aspects of topology, I.M.JAMES & E.H.KRONHEIMER (eds.)
94. Representations of general linear groups, G.D.JAMES
95. Low dimensional topology 1982: Proceedings of the Sussex Conference, 2-6 August 1982, R.A.FENN (ed.)
96. Diophantine equations over function fields, R.C.MASON
97. Varieties of constructive mathematics, D.S.BRIDGES & F.RICHMAN
98. Localization in Noetherian rings, A.V.JATEGAONKAR
99. Methods of differential geometry in algebraic topology, M.KAROUBI & C.LERUSTE
100. Stopping time techniques for analysts and probabilists, L.EGGHE

London Mathematical Society Lecture Note Series: 92

Representation of rings over skew fields

A.H. Schofield

Fellow of Trinity College, Cambridge

The right of the University of Cambridge to print and sell all manner of books was granted by Henry VIII in 1534. The University has printed and published continuously since 1584.

CAMBRIDGE UNIVERSITY PRESS

Cambridge

London New York New Rochelle

Melbourne Sydney

Published by the Press Syndicate of the University of Cambridge
The Pitt Building, Trumpington Street, Cambridge CB2 1RP
32 East 57th Street, New York, NY 100-2, USA
10 Stamford Road, Oakleigh, Melbourne 3166, Australia

First published 1985

Printed in Great Britain at the University Press, Cambridge

Library of Congress catalogue card number: 84-22996

British Library cataloguing in publication data

Schofield, A.H.

Representation of rings over skew fields.—
(London Mathematical Society lecture note
series; 92)
1. Rings (Algebra)
I. Title   II. Series
512'.4   QA247

ISBN 0 521 27853 8

CONTENTS

To Nicky

PREFACE

The finite dimensional representations of a ring over commutative fields have been studied in great detail for many types of ring, for example, group rings or the enveloping algebras of finite dimensional Lie algebras, but little is known about the finite dimensional representations of a ring over skew fields although such information might be of great use. The first part of this book is devoted to a classification of all possible finite dimensional representations of an arbitrary ring over skew fields in terms of simple linear data on the category of finitely presented modules over the ring. The second part is devoted to a fairly detailed study of those skew fields that arise in the first part and in the work of Cohn on firs and skew fields.

As has been said, the main goal at the beginning is to study finite dimensional representations of a ring over skew fields. An alternative view of this is that we should like to classify all possible homomorphisms from a ring to simple artinian rings; such a study was carried out in the case of one dimensional representations which are simply homomorphisms to skew fields by Cohn who showed that these homomorphisms are determined by which sets of matrices become zero-divisors over the skew field and gave a characterisation of the sets of matrices that could be exactly those that become singular under a homomorphism to a skew field. This theory has a particular application to firs, rings such that every left and right ideal are free of unique rank to show that they have universal homomorphisms to skew fields. This applies to the free algebra over a commutative field, and the ring coproduct of a family of skew fields amalgamating a common skew subfield, and gives the free skew field on a generating set and the skew field coproduct with amalgamation.

In order to classify homomorphisms from a ring to simple artinian rings, it is necessary to investigate what type of information a homomorphism gives on the ring. The most obvious point is that it induces certain rank functions on the modules over the ring; over a simple artinian ring, $S = M_n(D)$, where D is a skew field, every module is a direct sum of copies of the simple module, and the free module of rank one is the direct sum of n copies of the simple module, so we can assign a rank to the finitely generated modules over the ring taking values in $\frac{1}{n}\mathbb{Z}$ so that the free module on one generator has rank 1. If there is a homomorphism from R to S, we may assign ranks to the f.g. projectives or more generally the finitely presented modules over R by $\rho(M)$ is the rank over S of $M \otimes_R S$. Considering the rank functions induced on finitely generated projectives is important for constructing universal homomorphisms from an hereditary ring to simple artinian rings, whilst rank functions on finitely presented modules are precisely what is needed to classify all homomorphisms from the ring to simple artinian rings. The main result states that a rank function ρ on the finitely presented modules over a k-algebra R taking values in $\frac{1}{n}\mathbb{Z}$ arises from a homomorphism to a simple artinian ring if and only if it satisfies the following axioms:

1. $\rho(R^1) = 1$;
2. $\rho(A \oplus B) = \rho(A) + \rho(B)$;
3. if $A \to B \to C \to 0$ is an exact sequence of finitely presented modules then $\rho(C) \le \rho(B) \le \rho(A) + \rho(C)$.

If R is a ring that is not a k-algebra, it is necessary to have a fourth axiom:

4. $\rho(R/mR) = 0$ or 1 for any integer m.

Two homomorphisms $\alpha_i : R \to S_i$ induce the same rank function if and only if there is a commutative diagram of rings:

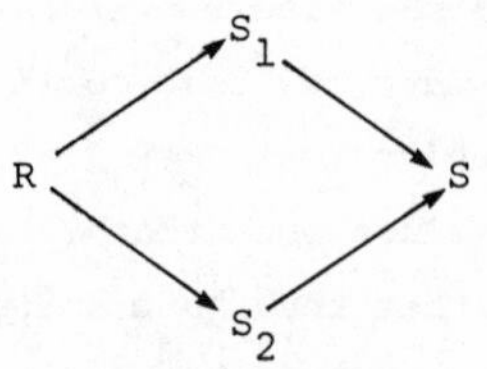

In chapter 1, we begin the study of hereditary rings and rank functions on finitely generated projectives over them. In the main, it is a study of the category of finitely generated projectives and the ranks that the rank function induces on the maps in the category. It is shown that this behaves in a very similar way to the rank functions on von Neumann regular rings, which is where the notion of a rank function came from; this analogy is developed to its logical conclusion in chapter 6, where it is shown that a rank function taking values in the real numbers defined on the f.g. projectives over an hereditary ring must arise from a homomorphism to a von Neumann regular ring.

Chapter 2 sets forth the first of the ring constructions that are needed in order to construct homomorphisms, the ring coproduct amalgamating a semisimple artinian subring. On the whole, it is a summary without proofs of Bergman's coproduct theorems. Chapter 3 shows how projective rank functions behave under the coproduct construction. It is also shown that if a module M over R_1 requires n generators then the module $M \otimes_R R'$ over the ring coproduct R' of R_1 and R_2 amalgamating a skew subfield F still requires n generators provided that there are finitely generated modules over R_2 requiring arbitrary large numbers of generators; the condition is clearly necessary. This may be regarded as the analogue of the Grushko Neumann theorem. The results on projective rank functions are applied to prove a recent theorem due to Linnell; a finitely generated group is accessible if there is a bound on the size of finite subgroups.

Chapter 4 presents the second important construction, adjoining universal inverses to maps between finitely generated projectives over a ring; this was studied by Cohn for matrices in order to construct homomorphisms to skew fields, but it has usually been regarded as a difficult technique, although it has arisen, usually in disguised form, in a number of contexts. For example, one of the methods used for showing that some finite dimensional algebra is of wild representation type amounts to adjoining a universal inverse to a suitable map. There are a number of ways of studying this construction developed recently which make it a little easier to calculate with and to think about, and the aim of this chapter is to present them. At the end, the algebraic K-theory of a universal localisation is discussed; there is an exact sequence for the algebraic K-theory that generalises the Bass, Murthy sequence for central localisation.

Chapter 5 pulls together the various pieces presented in the first four chapters in order to construct universal homomorphisms from an hereditary algebra with a rank function on its finitely generated projectives to a simple artinian ring. The idea is fairly simple; given a rank function ρ on an hereditary ring R, we ask which maps between finitely generated projectives have a chance of becoming invertible under a homomorphism from R to a simple artinian ring that induces the given rank function; if $\alpha : P \to Q$ is such a map, then $\rho(P) = \rho(Q)$ and α cannot factor through a projective of smaller rank. Such maps are called full maps. The universal localisation of an hereditary ring at all full maps with respect to a rank function taking values in $\frac{1}{n}\mathbb{Z}$ is a perfect hereditary ring and it is simple artinian in a large number of interesting cases. Chapter 6 completes this circle of ideas by showing that if R is an hereditary ring with a rank function on its finitely generated projectives taking values in the real numbers, there is a homomorphism from R to a von Neumann regular ring with a unique rank function that induces this rank function on R. This theorem actually holds provided that all countably generated right and left ideals over R are projective, which means that it applies to a von Neumann regular ring with a rank function.

Chapter 7 contains a number of results on homomorphisms to simple artinian rings beyond those that were discussed above. The space of all possible rank functions on finitely presented modules over a k-algebra that satisfy the axioms we stated earlier form in a natural way a $\mathbb{Q}$-convex subset of an infinite dimensional vector space. Given two rank functions that satisfy the axioms given, so does the rank function $q_1\rho_1 + q_2\rho_2$ where q_1 and q_2 are positive rationals such that $q_1 + q_2 = 1$. It is shown in the course of chapter 7 that every rank function ρ has a unique expression in the form $\Sigma q_i\rho_i$ where q_i are positive rationals such that $q_i = 1$ and ρ_i are rank functions that cannot be written as the weighted sum of different rank functions. So, the space of all possible rank functions is a sort of locally finite dimensional $\mathbb{Q}$-simplex.

The methods and theorems developed in the first part of this book are of great use in studying the skew fields constructed by Cohn, and the second half of this book is a fairly detailed investigation along these lines.

In chapter 8, we investigate what is known about the centre of the skew field and simple artinian coproduct. We have a complete answer when we amalgamate over a central subfield; however, the results are rather incomplete for simple artinian coproducts where none of the factors are skew fields. Chapter 9 continues with a detailed discussion of the finite dimensional division subalgebras of skew field coproducts and a number of other related skew fields. As an example of the odd results that occur, it is shown that if E_1 and E_2 are skew fields containing no elements algebraic over the central subfield k, then $E_1 \underset{k}{\circ} E_2$ the skew field coproduct of E_1 and E_2 amalgamating k can sometimes contain a finite dimensional field extension L of K, but if it does, $[L:k]$ must be divisible by two different primes; there is an example where $[L:k] = 6$. There is also an example of a skew field D with centre k such that $D\otimes_k k_s$ and $D\otimes_k kP^{-\infty}$ are skew fields where k_s is the separable closure of k, and $kP^{-\infty}$ is the inseparable closure of k but $D\otimes_k \bar{k} \cong M_p(D')$ for some skew field D', where $\bar{k}$ is the algebraic closure of k; this settles a question of Cohn and Dicks.

Chapter 10 develops the technique of the universal bimodule of derivations in order to distinguish between various non-isomorphic skew fields. In particular, it is shown that the free skew field on m generators cannot be isomorphic to the free skew field on n generators for $m \neq n$. It also gives a way for recognising when a skew field is a universal localisation of an hereditary subring.

Chapter 11 continues the investigation of the skew subfields of a skew field coproduct; we are particularly interested in the commutative subfields of such skew fields and in centralisers in matrix rings over a free skew field. In the first case, it is possible to bound the transcendence degree of commutative subfields of a skew field coproduct in terms of the transcendence degree of commutative subfields of the factors and the amalgamated skew field of the coproduct. For centralisers, it is shown that a skew subfield D with transcendental centre of $M_n(F)$ where F is a free skew field over k has a finitely generated centre over k of transcendence degree 1, its dimension over its centre is finite, and this dimension must divide n^2. At the end of the chapter, it is shown that a 2 generator skew subfield of a free skew field must either be free on those 2 generators or else it is commutative.

Chapter 12 develops the characterisation of the universal localisations of hereditary rings that are skew fields which was developed in chapter 10 into a characterisation of simple artinian universal localisations of hereditary rings; then it is shown that if T is a subring of a simple artinian universal localisation of the hereditary ring R that contains the image of R so that the map from R to T is an epimorphism, then T is itself a universal localisation of R. It follows from this result that epic endomorphisms of the free algebra over a commutative field are isomorphisms; this is the non-commutative analogue of the Jacobian conjecture.

The final chapter presents among other things a solution to an old problem; it is shown that for any pair of integers $a,b > 1$, there exists an extension of skew fields $E \supset F$ such that the left dimension of

E over F is a, whilst the right dimension is b. By extending the construction, it is possible to construct a new class of hereditary artinian rings of finite representation type. In order to effect these constructions, we develop a new type of hereditary ring construction, the bimodule amalgam rings; these are rings generated by two simple artinian rings S and S' subject only to conditions on the S, S' bimodule $SS' = \{\Sigma s_i s'_1 : s_i \in S, s'_i \in S'\}$. When we are able to show that these hereditary rings have a rank function, their properties are of particular interest. In addition to the results mentioned above, they also allow us to construct isomorphisms between skew fields that at first glance appear to be quite different. As an example, it is shown that if E_1 and E_2 are division subalgebras of the skew field F such that $[E_1:k] = [E_2:k]$ where k is a central subfield, then $F \circ_k E_1$ is isomorphic to $F \circ_k E_2$.

There are a number of people that I should like to thank for their encouragement and help during the proving of these results and subsequently during the time that I was writing them down. The first person I should like to thank is Warren Dicks with whom I have discussed most of the results of this book; his care and accuracy have been of great assistance to me and many of the results have arisen out of conversations between us. I should also like to thank Paul Cohn for his interest and encouragement; I owe him a particular debt for having proven the first results in this area. I should also like to thank Rufus Neal for bearing with me despite the length of time that it has taken me to get this book into his hands at Cambridge University Press, and I am very grateful to Diane Quarrie for typing this book so well from a partial typescript of poor quality.

PART I

Homomorphisms to simple artinian rings

1 HEREDITARY RINGS AND PROJECTIVE RANK FUNCTIONS

Definitions and preliminaries

In this chapter, we introduce the two main subjects of the first part of this book; hereditary rings and the projective rank functions on the rings, which we need in order to study their homomorphisms to simple artinian rings.

A left hereditary ring is one such that all left ideals are projective modules. We shall be interested in a number of variants of this definition; a left semihereditary ring is one such that all finitely generated left ideals are projective and a left $\aleph_0$-hereditary ring is one such that all countably generated left ideals are projective. We shall often need to consider the two-sided properties, whose definitions we leave to the reader; our results tend to work most often in the case of two-sided $\aleph_0$-hereditary rings. We shall tend to miss out the words 'two-sided', when using these conditions. There is a two-sided condition implied by all of the one-sided conditions above; a ring is weakly semihereditary if, for all pairs of maps $\alpha : P_0 \to P_1, \beta : P_1 \to P_2$ between finitely generated projective modules such that $\alpha\beta = 0$, then $P_1 = P_1' \oplus P_1''$, where the image of α lies in P_1' and the kernel of β contains P_1'. This is a two-sided condition because of the duality between the category of finitely generated left projectives and finitely generated right projectives induced by $\mathrm{Hom}_R(\_,R)$.

We shall abbreviate 'finitely generated' to f.g., 'finitely presented' to f.p., and, in the case of vector spaces over a skew field, we abbreviate 'finite dimensional' to f.d.

Lemma 1.1 A left semihereditary ring is weakly semihereditary.

Proof: Suppose that $\alpha : P_0 \to P_1, \beta : P_1 \to P_2$ are two maps such that $\alpha\beta = 0$, where P_i is a finitely generated projective for $i = 0, 1, 2$. The image of β is a projective module, so $P_1 = \mathrm{im}\beta \oplus \ker\beta$, and the image of α lies in the kernel of β.

A good reason for introducing the notion of a weakly semi-hereditary ring is the following theorem due to Bergman.

Theorem 1.2 Every projective module over a weakly semihereditary ring is a direct sum of finitely generated projective modules.

We refer the reader to 0.2.9 of Cohn (71) for a proof of this result.

We shall work as long as possible with weakly semihereditary rings; however, we shall eventually be forced to restrict our attention to two-sided $\aleph_0$-hereditary rings. This class of rings draws much of its initial interest from the fact that all von Neumann regular rings have this property. By a von Neumann regular ring, we mean a ring R, such that for all x in R, there exists an element y such that $xyx = x$; we shall see that there are interesting connections between these classes of rings.

Much of the work of this chapter is just a study of the category of finitely generated left projective modules over a weakly semihereditary ring. This has been done with a great deal of success for semifirs and firs by Cohn (71); a fir is a ring such that all left ideals and right ideals are free of unique rank, and a semifir is a ring such that all finitely generated left ideals (and so, all such right ideals too) are free of unique rank. In this case, the arguments work well because we have a good notion of the size of a finitely generated projective, and so we would like to have a generalisation of this idea for other rings. The relevant idea comes from the theory of von Neumann regular rings.

Given a ring, R, we associate to it the abelian monoid $P_{\oplus}(R)$ of isomorphism classes of f.g. projectives under direct sum. We may also associate to it a pre-ordered abelian group, the Grothendieck group, $K_0(R)$. It is generated by the isomorphism classes of finitely generated left projective modules $[P]$, subject to the relations $[P\oplus Q] = [P] + [Q]$, for every pair of isomorphism classes $[P]$, $[Q]$. The pre-order is given by specifying a positive cone, by which we mean simply a distinguished additive submonoid of positive elements, and, in this case we take the isomorphism classes of finitely generated projective modules, $[P]$. It is clear that $K_0(R)$ is the universal group associated to $P(R)$. Two projectives P and Q are said to be stably isomorphic when $[P] = [Q]$; this is equivalent to the existence of an equation $P \oplus R^n \cong Q \oplus R^n$.

A projective rank function on a ring R is a homomorphism of pre-ordered groups, $\rho : K_0(R) \to \mathbb{R}$, the real numbers, such that $\rho([R^1]) = 1$. By definition, $\rho([P]) \geq 0$; we shall call a rank function faithful if $\rho([P]) > 0$, for all non-zero P. We shall often simplify the notation by writing $\rho(P)$ for $\rho([P])$. We note that a projective rank function is a left, right dual notion, because of the duality $\mathrm{Hom}_R(-,R)$.

A partial projective rank function is a homomorphism of pre-ordered groups $\rho : A \to \mathbb{R}$, where A is a subgroup of $K_0(R)$, containing $[R^1]$, and the partial order is that induced from $K_0(R)$ by restriction.

We recall theorem 18.1 of Goodearl (79):

Theorem 1.3 Every partial projective rank function extends to a projective rank function on R.

This result allows us to characterise those rings that have a projective rank function. We say that a ring has unbounded generating number if for every natural number n, there is a finitely generated module, M, requiring at least n generators. It is an easy check that this equivalent to the condition that for no m is there an equation of the form $R^m \cong R^{(m+1)} \oplus P$; and this is a left, right dual condition, which justifies the two-sided nature of our definition. Cohn mentions this class of rings in (Cohn 71) under the guise of rings such that for all n, the n by n identity matrix cannot be written as an n by $(n-1)$ matrix times an $(n-1)$ by n matrix. We leave it to the reader to check the equivalence.

Theorem 1.4 A ring has a projective rank function, if and only if it has unbounded generating number.

Proof: Certainly, if R has a projective rank function, it must have unbounded generating number.

Conversely, if R has unbounded generating number, the subgroup of $K_0(R)$ generated by $[R^1]$ is isomorphic to $\mathbb{Z}$, and under the isomorphism, no stably free projective module can have negative image. So this isomorphism defines a partial projective rank function on R, which must extend to a projective rank function by 1.3.

We have shown that most rings have a projective rank function; in fact, projective rank functions arise quite naturally on rings and one is forced to study them in order to solve certain types of problems.

If S is a simple artinian ring, it has the form $M_n(D)$ for some skew field D, and so, $K_0(S)$ can be identified in a natural way with $\frac{1}{n}\mathbb{Z}$; so in this case, we have a unique rank function. If we have a homomorphism from a ring R to S, this induces a homomorphism from $K_0(R)$ to $K_0(S)$, which is naturally isomorphic to $\frac{1}{n}\mathbb{Z}$; therefore, homomorphisms to simple artinian rings induce rank functions to $\frac{1}{n}\mathbb{Z}$, and we shall need to consider such rank functions in order to study homomorphisms to simple artinian rings. More generally, many von Neumann regular rings have rank functions so that in order to study homomorphisms to von Neumann regular rings we shall need to consider quite general projective rank functions.

These projective rank functions appear naturally in the representation theory of finite dimensional algebras, for if R is a finite dimensional algebra over the field k, and M is a finite dimensional module, $[M:k] = m$, this defines a homomorphism from R to $M_m(k)$ and so determines a rank function ρ on R; it is easy to see that if P is a principal projective module over R, $P = Re$, then

$$\rho(P) = \frac{[Me:k]}{[M:k]}$$

which determines the projective rank function, since all f.g. projective modules are direct sums of principal projective modules for an artinian ring.

Another class of rings with a projective rank function that occurs naturally are the group rings in characteristic 0. We have a trace function on the group ring FG, where F is a field of characteristic 0 and G is a group given by $tr(\sum_i f_i g_i) = f_0$, where g_0 is the identity element of the group. We extend this to a trace function on the ring $M_n(FG)$ in the natural way and then we define the rank of an f.g. projective P to be the trace of an idempotent e in $M_n(FG)$ such that $FG^n e \cong P$. It is well known that this is well-defined, taking values in $\mathbb{Q}$, and that it is a faithful projective rank function. This will turn out to be useful to us later on in proving results due to Linnell on accessibility of f.g. groups.

<u>Trace ideals</u>

It is often useful to be able to work with a faithful rank function on a ring rather than one that is not; so we should like to have a way of getting rid of the projectives of rank zero. There is a standard way of dealing with this problem; we define the trace ideal of a set of f.g. projective modules, I, closed under direct sum to be the set of elements, T, that lie in the image of some map from one of these projectives to the free module of rank 1. It is easy to see that this set is an ideal, in fact, an idempotent ideal known as the trace ideal of the projectives in I, and that R/T is the universal R-ring such that $R/T \otimes_R P = 0$, for all P in I. We wish to study the behaviour of this construction.

Theorem 1.5 Let I be a set of f.g. projective modules closed under direct sum over a ring R and let T be the trace ideal of this set of projectives. For a f.g. projective, Q, $R/T \otimes_R \ = 0$, if and only if Q is a direct summand of an element of I. The monoid of induced projective modules over R/T is the quotient of $P_{\oplus}(R)$ by the relation $P \sim P'$, if and only if $P \oplus Q \cong P' \oplus Q'$, where Q and Q' are direct summands of elements of I.

Proof: Suppose that $R/T \otimes_R Q = 0$, then every element of Q lies in the image of a map from an element of I to Q; since Q is finitely generated, there must be a surjective map from an element of I to Q, which proves the first assertion.

Suppose that $R/T \otimes_R \alpha : R/T \otimes_R P \to R/T \otimes_R P'$ is an isomorphism over R/T; so there is a surjection:
$\alpha \oplus \beta : P \oplus Q \to P'$, where Q is a direct summand of an element of I. Therefore, $0 \to \ker \alpha\oplus\beta \to P \oplus Q \to P' \to 0$ is a split exact sequence, where $R/T \otimes_R \alpha \oplus \beta$ is an isomorphism, since it equals $R/T \otimes_R \alpha$. So $\ker\alpha \oplus \beta$ becomes 0 over R/T, and must be a direct summand of an element of I. Therefore, as required, we have an equation of the form $P \oplus Q \cong P' \oplus Q'$. The converse is clear.

We have the following consequence:

Theorem 1.6 Let R be a ring with a projective rank function ρ; let T be the trace ideal of the projectives of rank 0; then ρ extends to a projective rank function on R/T.

Proof: By the last theorem, there is a partial projective rank function defined on the image of $K_0(R)$ in $K_0(R/T)$, induced by ρ. By theorem 1.3, this extends to a rank function on R/T.

We can do rather better than this on a weakly semihereditary ring. First, we need the following result on the behaviour of $P_\oplus$ on passing to the quotient by a trace ideal over a weakly semihereditary ring.

Theorem 1.7 Let R be a weakly semihereditary ring, and let T be the trace ideal of the set of f.g. projectives, I, closed under direct sum; then R/T is a weakly semihereditary ring, and $P_\oplus(R/T)$ is the quotient of $P_\oplus(R)$ by the relation $P \sim P'$ if and only if $P \oplus Q \cong P' \oplus Q'$, where Q and Q' are direct summands of an element of I.

Proof: We denote passage to R/T by bars, so $\bar{R} = R/T$.

Let $\bar{\alpha} : \bar{P} \to \bar{P}', \beta : \bar{P}' \to \bar{P}''$ be a pair of maps such that $\bar{\alpha}\bar{\beta} = 0$; then over R, $\alpha\beta = \gamma\delta, \gamma : P \to Q, \delta : Q \to P''$, where Q is an element of I. So, over R, we have

$$(\alpha\ \gamma) \begin{pmatrix} \beta \\ -\delta \end{pmatrix} = 0$$

and we note that $\overline{(\alpha\gamma)} = \bar{\alpha}$, and $\begin{pmatrix} \bar{\beta} \\ -\bar{\delta} \end{pmatrix} = \bar{\beta}$.

Since R is weakly semihereditary, $P' \oplus Q \cong P_1 \oplus P_2$, where $\mathrm{im}(\alpha\gamma) \subseteq P_1$, and $\ker \begin{pmatrix} \beta \\ -\delta \end{pmatrix} \supseteq P_1$; therefore, $\bar{P}' \cong \bar{P}_1 \oplus \bar{P}_2$, where $\mathrm{im}\bar{\alpha} \subseteq P_1 \subseteq \ker \bar{\beta}$, that is, the weakly semihereditary condition is satisfied for maps between induced projectives.

Let $e : R^n \to R^n$ be a map such that $\bar{e}^2 = \bar{e}$; that is, $\bar{e}(\overline{1-e}) = 0$. So, by the previous argument, $R^n \cong P_1 \oplus P_2$, where $\mathrm{im}\ \bar{e} \subseteq \bar{P}_1 \subseteq \ker(\overline{1-e})$; but the image of $\bar{e}$ is equal to the kernel of $(\overline{1-e})$, so $\mathrm{im}\ \bar{e} = \bar{P}_1$, which shows that all f.g. projectives are induced, and, in consequence, R/T is weakly semihereditary.

The rest follows from theorem 1.5.

This allows us to pass from a weakly semihereditary ring with a projective rank function to a weakly semihereditary ring with a faithful projective rank function, simply by killing the projectives of rank 0. We summarise this special case:

Theorem 1.8 Let R be a weakly semihereditary ring with a projective rank function ρ; let T be the trace ideal of the f.g. projectives of rank 0; then R/T is a weakly semihereditary ring, and ρ induces a faithful projective rank function on R/T. If ρ takes values in $\frac{1}{n}\mathbb{Z}$, then R/T is semihereditary on either side.

Proof: All is clear except for the last remark. In this case, R/T is a weakly semihereditary ring with a faithful projective rank function, taking values in $\frac{1}{n}\mathbb{Z}$. Let M be a finitely generated left ideal and let P be a f.g. projective over R/T of minimal rank such that there is a surjection $\alpha : P \to M$. If x lies in the kernel of this surjection, we have a sequence:

$$R/T \to P \to M \subseteq R,$$

whose composite is 0; so $P \cong P_1 \oplus P_2$, where x is in P_1, which is in the kernel of the surjection. Hence, $\alpha|P_2 : P_2 \to M$, is a surjection. Since $\rho(P_2) < \rho(P)$ unless $x = 0$, we deduce that $\alpha : P \to M$ must be an isomorphism.

We have already noted that trace ideals must be idempotent; curiously, the converse is true for left hereditary rings as we see next.

<u>Theorem 1.9</u> Let R be a left hereditary ring, and let I be an idempotent ideal; then I is a trace ideal.

Proof: As a left module over R, I is projective and since $I = I^2$, $R/I \otimes_R I = I/I^2$. Hence, the trace ideal of the projective module I must contain I, but it can be no larger, since its image in R/I is trivial.

<u>The inner projective rank</u>

If we have a partial projective rank function, ρ_A, defined on the subgroup A of $K_0(R)$, we define the <u>generating number</u> with respect to ρ_A of a finitely generated left module M over R by the formula:

$$g.p_A(M) = \inf_{[P]}\{\rho_A(P): [P] \text{ is in } A, \exists \text{ a surjection } P \to M\} .$$

If all stably free modules are free of unique rank, and A' is the subgroup of $K_0(R)$ generated by $[R^1]$, the generating number with respect to $\rho_{A'}$, where $\rho_{A'}$ is the unique rank function defined on A is the minimal number of generators of a module. So we could hope and we shall show that our more general notion is a useful refinement.

We intend to use a projective rank function ρ to analyse the category of f.g. projectives over a ring R. We have a rank associated to each object of the category, so, our next aim is to give each map a rank. Let $\alpha : P \to Q$ be a map between two f.g. projectives; we define the <u>inner projective rank</u> of the map with respect to ρ to be given by the formula:

$$\rho(\alpha) = \inf_{[P']}\{\rho(P') : \exists \text{ a commutative diagram } \begin{array}{ccc} P & \rightarrow & Q \\ & \searrow \nearrow & \\ & P' & \end{array}\}$$

It is sometimes useful to have a related notion to hand; we define the <u>left nullity</u> of α to be $\rho(P) - \rho(\alpha)$. Similarly, the <u>right nullity</u> of α is defined by $\rho(Q) - \rho(\alpha)$. $\rho(P) = \rho(Q)$, the <u>nullity</u> of α is $\rho(P) - \rho(\alpha)$.

We may relate the inner projective rank of a map and the generating number of suitable modules.

<u>Lemma 1.10</u> Let R be a ring with a projective rank function ρ; then the inner rank of a map $\alpha : P \to Q$ is equal to the following: $\inf_M \{g.\rho(M) : \alpha(P) \subseteq M \subseteq Q$, where M is a f.g. submodule of $Q\}$.

In particular, if R is a left semihereditary ring,

$$\rho(\alpha) = \inf_{P'}\{\rho(P') : \alpha(P) \subseteq P' \subseteq Q\}.$$

Proof: If there is a commutative diagram $\begin{array}{ccc} P & \rightarrow & Q \\ & \searrow \nearrow & \\ & P' & \end{array}$, then $\alpha(P) \subseteq \beta(P') \subseteq Q$ and $g.\rho(\beta(P')) \le \rho(P')$, so $\rho(\alpha) \ge \inf_M\{g.\rho(M) : \alpha(P) \subseteq M \subseteq Q\}$.

Conversely, if $\alpha(P) \subseteq M \subseteq Q$, and there exists a surjection $P' \to M$, we have a commutative diagram $\begin{array}{ccc} P & \rightarrow & Q \\ & \searrow \nearrow & \\ & P' & \end{array}$, since P is projective.

Hence, $\rho(\alpha) = \inf_M\{g.\rho(M) : \alpha(P) \subseteq M \subseteq Q\}$.

A map $\alpha : P \to Q$ is said to be <u>left full</u> with respect to ρ if $\rho(\alpha) = \rho(P)$, and it is <u>right full</u> if $\rho(\alpha) = \rho(Q)$; it is <u>full</u> with respect to ρ if it is left and right full. The reason for considering full maps with respect to a projective rank function is that the only maps to have a chance of becoming inverted under a homomorphism to a simple artinian ring are the full maps with respect to the induced projective rank function; of course, it is in general rather unlikely that they all do; however, we shall find that for a hereditary ring there are homomorphisms for each projective rank function that invert all full maps.

We see that entirely the same theory may be set up on the dual category of f.g. right projectives, where the rank of a f.g. right projective module is that of its dual module. It is clear that the dual of a left full map is right full and vice versa.

The important fact about the inner projective rank on a weakly semihereditary ring is an analogue of Sylvester's law of nullity. We say that a ring satisfies the law of nullity with respect to ρ, or, alternatively, that the projective rank function ρ is a Sylvester projective rank function, if for every pair of maps between f.g. projectives $\alpha : P_0 \to P_1$, $\beta : P_1 \to P_2$ such that $\alpha\beta = 0$, then $\rho(\alpha) + \rho(\beta) \leq \rho(P_1)$. If R is a ring such that all f.g. projectives are free of unique rank, and this rank is a Sylvester projective rank function, R is a Sylvester domain.

Theorem 1.11 Let R be a weakly semihereditary ring with a rank function ρ; then ρ is a Sylvester projective rank function.

Proof: Recall that if $\alpha\beta = 0$ for $\alpha : P_0 \to P_1$, and $\beta : P_1 \to P_2$ over a weakly semihereditary ring, then $P_1 \cong P' \oplus P''$, where the image of α lies in P', and the kernel of β contains P', so that β factors through P''. Hence $\rho(\alpha) \leq \rho(P')$, and $\rho(\beta) \leq \rho(P'')$, so that $\rho(\alpha) + \rho(\beta) \leq \rho(P_1)$.

There are a few results that we can deduce from the law of nullity for a projective rank function on a ring. On the whole, they are a little technical, but since we shall need them later, it seems better to bore the reader now than to break up the flow of later proofs. Their point is to demonstrate the analogy between these rings with Sylvester projective rank functions and simple artinian rings with the standard rank function.

Lemma 1.12 Let R be a ring with a Sylvester projective rank function ρ; then for any pair of maps $\alpha : P_0 \to P_1$, $\beta : P_1 \to P_2$, $\rho(\alpha\beta) \geq \rho(\alpha) + \rho(\beta) - \rho(P_1)$. In particular, this holds for weakly semihereditary rings for any projective rank function.

Proof: Suppose that

$$\begin{array}{ccc} P_0 & \xrightarrow{\alpha\beta} & P_2 \\ & {\scriptstyle\gamma}\searrow \quad \nearrow{\scriptstyle\delta} & \\ & Q & \end{array}$$

is a commutative diagram. Then we have the maps $(\alpha|\gamma) : P_0 \to P_1 \oplus Q$, $\left(\frac{\beta}{-\delta}\right) : P_1 \oplus Q \to P_2$, and $(\alpha|\gamma)\left(\frac{\beta}{-\delta}\right) = 0$;

so, by the law of nullity, $\rho(\alpha|\gamma) + \rho\begin{pmatrix}\beta \\ -\delta\end{pmatrix} \le \rho(P_1) + \rho(Q)$. But $\rho(\alpha) \le \rho(\alpha|\gamma)$ and $\rho(\beta) \le \rho\begin{pmatrix}\beta \\ -\delta\end{pmatrix}$; therefore $\rho(\alpha) + \rho(\beta) - \rho(P_1) \le \rho(Q)$

Taking the infimum on the right proves the lemma.

We have the following corollary:

Corollary 1.13 Let R be a ring with a Sylvester projective rank function ρ. If α is right full, then $\rho(\alpha\beta) = \rho(\beta)$; dually, if α is left full, then $\rho(\beta\alpha) = \rho(\beta)$. In particular, the composite of left full maps is left full, and the composite of right full maps is right full.

Proof: Let $\alpha : P_0 \to P_1$ be right full, and let $\beta : P_1 \to P_2$ be some map; then $\rho(\beta) \ge \rho(\alpha\beta) \ge \rho(\beta) + \rho(\alpha) - \rho(P_1) = \rho(\beta)$. So, $\rho(\alpha\beta) = \rho(\beta)$. Therefore, the composite of right full maps is right full. The rest follows by duality.

Lemma 1.14 Let R be a ring with a Sylvester projective rank function ρ. Let $\alpha : P_1 \to P_2$, and $\beta : Q_1 \to Q_2$ be a pair of maps; then

$$\rho(\alpha) + \rho(\beta) \le \rho\begin{pmatrix}\alpha & 0 \\ \gamma & \beta\end{pmatrix} \le \rho(\alpha) + \rho(Q_1),\ \rho(\beta) + \rho(P_2).$$

Proof: Let P be a f.g. projective through which $\begin{pmatrix}\alpha & 0 \\ \gamma & \beta\end{pmatrix}$ factors; then we have an equation

$$\begin{pmatrix}\alpha & 0 \\ \gamma & \beta\end{pmatrix} = \begin{pmatrix}\delta_1 \\ \delta_2\end{pmatrix}(\varepsilon_1 \quad \varepsilon_2)$$

Since $\delta_1\varepsilon_2 = 0$, the law of nullity shows that $\rho(\delta_1) + \rho(\varepsilon_2) \le \rho(P)$; so $\rho(\alpha) + \rho(\beta) \le \rho(\delta_1) + \rho(\varepsilon_2) \le \rho(P)$, and taking the infimum on the right shows the first inequality.

If $\alpha = \delta\varepsilon$, $\delta : P_1 \to P$, $\varepsilon : P \to P_2$, then $\begin{pmatrix}\alpha & 0 \\ \gamma & \beta\end{pmatrix} = \begin{pmatrix}\delta & 0 \\ 0 & I\end{pmatrix}\begin{pmatrix}\varepsilon & 0 \\ \gamma & \beta\end{pmatrix}$; so $\rho\begin{pmatrix}\alpha & 0 \\ \gamma & \beta\end{pmatrix} \le \rho(P) + \rho(Q_1)$, and taking the infimum over $\rho(P)$ shows that

$$\rho\begin{pmatrix}\alpha & 0 \\ \gamma & \beta\end{pmatrix} \le \rho(\alpha) + \rho(Q_1) ;$$

similarly, if $\beta = \delta'\varepsilon'$, $\delta' : Q_1 \to Q$, $\varepsilon' : Q \to Q_2$, then we have the equation:

$$\begin{pmatrix}\alpha & O\\ \gamma & \beta\end{pmatrix} = \begin{pmatrix}\alpha & O\\ \gamma & \delta'\end{pmatrix}\begin{pmatrix}I & O\\ O & \varepsilon'\end{pmatrix} .$$

So $\rho\begin{pmatrix}\alpha & O\\ \gamma & \beta\end{pmatrix} \leq \rho(Q) + \rho(P_2)$, and taking the infimum over $\rho(Q)$, we see that

$$\rho\begin{pmatrix}\alpha & O\\ \gamma & \beta\end{pmatrix} \leq \rho(\beta) + \rho(P_2) .$$

We have one more dull lemma to put behind us:

Lemma 1.15 Let R be a ring with a Sylvester projective rank function ρ; if α is right full or β is left full, $\rho\begin{pmatrix}\alpha & O\\ \gamma & \beta\end{pmatrix} = \rho(\alpha) + \rho(\beta)$. Also, for all α and β, $\rho\begin{pmatrix}\alpha & O\\ O & \beta\end{pmatrix} = \rho(\alpha) + \rho(\beta)$.

Proof: We use the notation of the last lemma.

$\rho(\alpha) + \rho(\beta) \leq \rho\begin{pmatrix}\alpha & O\\ \gamma & \beta\end{pmatrix} \leq \rho(\alpha) + \rho(Q_1)$; if β is left full, $\rho(\beta) = \rho(Q_1)$, and so $\rho(\alpha) + \rho(\beta) = \rho\begin{pmatrix}\alpha O\\ \gamma\beta\end{pmatrix}$. A similar argument works if α is right full.

If $\alpha = \delta_1\delta_2$, and $\beta = \varepsilon_1\varepsilon_2$, we have the equation:

$$\begin{pmatrix}\alpha & O\\ O & \beta\end{pmatrix} = \begin{pmatrix}\delta_1 & O\\ O & \delta_2\end{pmatrix}\begin{pmatrix}\varepsilon_1 & O\\ O & \varepsilon_2\end{pmatrix}$$

It follows that $\rho\begin{pmatrix}\alpha & O\\ O & \beta\end{pmatrix} = \rho(\alpha) + \rho(\beta)$.

In order to get fairly decisive results, it is necessary to restrict our attention to a two-sided X_o-hereditary ring with a faithful rank function. Dicks pointed out that the proof of the next main theorem, originally stated for two-sided hereditary rings actually holds in the greater generality. It is the central point in the proof that a rank function on a two-sided X_o-hereditary algebra arises from a homomorphism to a von Neumann regular ring, and the reader may well wish to read it only when he comes to this theorem in chapter 6.

Theorem 1.16 Let R be a two-sided X_o-hereditary ring with a faithful projective rank function. Then every map between f.g. projectives factors as a right full followed by a left full map.

Proof: First, we show that every map has a non-trivial right factor that is left full.

If $\alpha : P \to Q$ is not left full or right full, then, by Lemma 1.10, there exists P_1, $\operatorname{im}(\alpha) \subseteq P_1 \subseteq Q$, such that $\rho(P_1) < \rho(P), \rho(Q)$, and $\rho(P_1) - \rho(\alpha) < 1$.

If $\alpha_1 : P_1 \to Q$ is not a left full embedding, we may find P_2 such that $P_1 \subseteq P_2 \subseteq Q$, $\rho(P_2) < \rho(P_1)$, and $\rho(P_2) - \rho(\alpha_1) < \frac{1}{2}$; in general, at the nth stage, if $\alpha_{n-1} : P_{n-1} \to Q$ is not a left full map, we choose P_n, $P_{n-1} \subseteq P_n \subseteq Q$, such that $\rho(P_n) < \rho(P_{n-1})$, and $\rho(P_n) - \rho(\alpha_{n-1}) < \frac{1}{n}$; if this process does not terminate, we obtain the chain:

$$\operatorname{im}\alpha \subset P_1 \subset P_2 \subset \ldots\ldots\ldots\ldots \subset P_n \subset \ldots\ldots \subset Q .$$

$\bigcup_n P_n$ is a countably generated submodule of Q, so it is projective, and must be a direct sum of f.g. projectives by 1.2. So $\bigcup P_n \cong P_1' \oplus P_2' \oplus \ldots$, where each P_i' is a f.g. projective module.

Since $\operatorname{im}\alpha$ is finitely generated, it lies in $P_N'' = \bigoplus_{i=1}^{N} P'$, for some N. We claim that the embedding $P_N'' \subseteq Q$ is a left full map.

Since P_N'' is finitely generated, $P_N'' \subseteq P_n$ for some n, and so, $P_N'' \subseteq P_m$ for all $m \geq n$. Moreover, it is a direct summand of each such P_m, since it is a direct summand of their union.

Consider a f.g. submodule, Q', such that $P_N'' \subseteq Q' \subseteq Q$; consider the split exact sequence:

$$0 \to P_m \cap Q' \to P_m \oplus Q' \to P_m + Q' \to 0 \quad ;$$

$P_m \cap Q' \supseteq P_N''$, so $P_m \cap Q' \cong P_N'' \oplus Q''$, for some module Q'', and so $\rho(P_m \cap Q') \geq \rho(P_N'')$.

Also, $P_{m-1} \subset P_m \subseteq P_m + Q'$, so that $\rho(P_m + Q') \geq \rho(P_m) - \frac{1}{m}$.
From the exact sequence, $\rho(P_m) + \rho(Q') = \rho(P_m + Q') + \rho(P_m \cap Q')$, and we have just shown that the right hand side is greater than or equal to $\rho(P_N'') + \rho(P_m) - \frac{1}{m}$, so that we can deduce $\rho(Q') \geq \rho(P_N'') - \frac{1}{M}$; but this holds for all $m \geq n$, and so we conclude that $P_N'' \subseteq Q$ must be a left full map that is a right factor of $\alpha : P \to Q$; by construction, $\rho(P_N'') < \rho(Q)$.

By duality, we deduce that we can always find a right full left factor of $\alpha : P \to Q$, when α is not left or right full, $\alpha = \beta_1\gamma$, where $\beta_1 : P \to Q_1$, where $\rho(Q_1) < \rho(P)$. Since a sequence of right full maps is right full, by 1.13, we may assume that the right full map is to some submodule Q_1 of Q. So we have $\mathrm{im}\alpha \subseteq Q_1 \subseteq Q$, where the induced map $\alpha' : P \to Q_1$ is right full. If the map $\gamma_i : Q_1 \subset Q$ is not left full, we can find Q_2, $Q_1 \subset Q_2 \subset Q$ such that $\rho(Q_2) < \rho(Q_2)$, and $Q_1 \subset Q_2$ is a right full map. We assume that this process does not finish and obtain a contradiction. If it continues, we have an ascending chain:

$$\mathrm{im}\alpha \subset Q_1 \subset Q_2 \subset \ldots. \subset Q, \quad \text{where} \quad \rho(Q_{i+1)} < \rho(Q_i) \quad .$$

$\cup_i Q_i$ is a countably generated submodule of Q, so it is projective and, by 1.2, $\cup_i Q_i \cong \oplus_j Q_j'$, where each Q_j' is f.g. projective. For some m, $Q_1 \subseteq \oplus_{j=1}^m Q_j'$; again, for some n, $\oplus_{j=1}^m Q_j' \subseteq Q_n \subset Q_{n+1}$, and $\oplus_{j=1}^m Q_j'$ is a direct summand of Q_{n+1}, so that we have $\rho(\oplus_{j=1}^m Q_j') \le \rho(Q_{n+1}) < \rho(Q_n)$. Since $Q_1 \subset Q_n$ factors through $\oplus_{j=1}^m Q_j'$, it cannot be right full, but it is the composite of the right full maps $Q_i \subset Q_{i+1}$, so that it is right full by 1.13. This contradiction shows that our process must end at a finite stage; that is, $Q_m \subset Q$ is left full for some m. But $P \to Q_m$ is the composite of right full maps, and must be right full. So, we have a factorisation of $\alpha : P \to Q$ as a right full followed by a left full map.

When every map factors as a right full then a left full map with respect to a projective rank function, we shall say that the ring, R, has <u>enough right and left full maps with respect to the projective rank function.</u> We describe a factorisation of a map as a right full by a left full map as <u>a minimal factorisation.</u>

<u>Corollary 1.17</u> Let R be a two-sided $\aleph_0$-hereditary ring with a faithful projective rank function ρ. Then for any map between f.g. projectives $\alpha : P \to Q$, $\rho(\alpha) = \min_{P'} \{\rho(P') : \mathrm{im}\alpha \subseteq P' \subseteq Q\}$.

Proof: This follows at once from 1.16. For $\alpha = \alpha_1\alpha_2$, where $\alpha : P \to P'$ is right full and α_2 is left full. By 1.12, $\rho(\alpha) \ge \rho(\alpha_1) + \rho(\alpha_2) - \rho(P')$; so $\rho(\alpha) = \rho(P')$.

There are occasions when this result is automatically true for a weakly semihereditary ring; for example, if we have the descending chain condition on the numbers $\rho(P)$. It fails in general for weakly semihereditary rings.

<u>The torsion modules for a projective rank function</u>

We pass from the study of quite general maps in the category of f.g. projectives to the study of the full maps with respect to a projective rank function. We shall assume that the projective rank function is faithful throughout this section. The sensible way to study the full maps with respect to a faithful Sylvester projective rank function is to look at the full subcategory of modules that are cokernels of full maps with respect to the projective rank function; we shall call this the category of <u>torsion left modules</u> with respect to the rank function. If $\operatorname{coker} \alpha \cong \operatorname{coker} \beta$, Schanuel's lemma shows that α and β are <u>stably associated</u>, that is, we have a matrices of maps equation:

$$\gamma \begin{pmatrix} \alpha & O \\ O & I_\rho \end{pmatrix} = \begin{pmatrix} \beta & O \\ O & I \end{pmatrix} \delta \qquad ,$$

where γ and δ are invertible maps.

<u>Theorem 1.18</u> Let R be a semihereditary ring with a faithful projective rank function ρ; then, <u>T</u>, the category of torsion modules with respect to ρ, is an abelian category.

Proof: Let M_1 and M_2 be torsion modules with presentations: $O \to Q_i \xrightarrow{\alpha_i} P_i \to M_i \to O$, where α_i is a full map, and let $\emptyset : M_1 \to M_2$ be some map.

We have a commutative diagram:

$$\begin{array}{ccccccccc} O & \to & Q_1 & \to & P_1 & \to & M_1 & \to & O \\ & & \downarrow & & \downarrow \emptyset' & & \downarrow \emptyset & & \\ O & \to & Q_2 & \to & P_2 & \to & M_2 & \to & O \end{array}$$

We have two presentations of $\operatorname{im}\emptyset$:

$$O \to Q_2 \to \operatorname{im}\emptyset' + Q_2 \to \operatorname{im}\emptyset \to O$$

$$O \to Q_1' \to P_1 \to \operatorname{im}\emptyset \to O \qquad ,$$

where $P_2' = \text{im}\emptyset' + Q_2$ is finitely generated, so it is projective, and Schanuel's lemma shows that $Q_2 \oplus P_1 \cong Q_1' \oplus P_2'$, so Q_1' is also f.g. projective. $Q_1 \subseteq Q_1' \subseteq P_1$, and $Q_2 \subseteq P_2' \subseteq P_2$, so that $\rho(Q_1') \geq \rho(Q_1)$, and $\rho(P_2') \geq \rho(Q_2)$; on putting this into $Q_2 \oplus P_1 \cong Q_1' \oplus P_2'$, we see that equality must hold; moreover, any f.g. module between Q_1' and P_1 lies between Q_1 and P_1 and so has rank at least $\rho(Q_1') = \rho(P_1)$, so $\text{im}\emptyset$ is a torsion module with respect to ρ. The kernel of $\emptyset$ is Q_1'/Q_1, and is also torsion, whilst coker $\emptyset \cong P_2/P'$ which is torsion. Finally, finite direct sums of torsion modules are torsion, which completes the proof that $\underline{T}$ is an abelian category.

Clearly, this result will have consequences for the factorisations of full maps. In particular, we shall need the following lemma later on.

<u>Lemma 1.19</u> Let R be a semihereditary ring with a faithful projective rank function ρ. Let $\alpha\beta = \begin{pmatrix} \gamma_1 & & 0 \\ & \ddots & \\ & & \gamma_n \end{pmatrix}$ where α, β and γ_i, $i = 1$ to n, are all full maps with respect to ρ; then there exists an invertible map $\emptyset$ such that

$$\alpha\emptyset = \begin{pmatrix} \alpha_1 & & 0 \\ & \ddots & \\ & & \alpha_n \end{pmatrix}, \quad \emptyset^{-1}\beta = \begin{pmatrix} \beta_1 & & 0 \\ & \ddots & \\ & & \beta_n \end{pmatrix}$$

where α_i and β_i are full maps such that $\alpha_i\beta_i = \gamma_i$.

Proof: It is sufficient to prove this for $n = 2$ since the general case follows by induction. So, consider $\alpha\beta = \begin{pmatrix} \gamma_1 & 0 \\ \delta & \gamma_2 \end{pmatrix}$; coker γ_1 embeds in coker $\alpha\beta$ with quotient module isomorphic to coker γ_2; coker $\alpha\beta$ maps onto coker α, and the image of coker γ_1 in coker α is a torsion module of the form coker α_1, where α_1 is a left factor of γ_1, whilst the quotient of coker α by α coker α_1 is a torsion module of the form coker α_2 where α_2 is a left factor of γ_2; therefore, there is an invertible map $\emptyset$ such that $\alpha\emptyset = \begin{pmatrix} \alpha_1 & 0 \\ \varepsilon_1 & \alpha_2 \end{pmatrix}$; inspection shows that $\emptyset^{-1}B$ takes the form $\begin{pmatrix} \beta_1 & 0 \\ \varepsilon_2 & \beta_2 \end{pmatrix}$.

In certain cases, we can show that this category of torsion

modules must be artinian and noetherian, which allows us to apply the Jordan-Hölder theorem in order to deduce the unique factorisation of maps into atomic full maps. Our method of proving this is to show that the category is noetherian in certain cases and then use a duality between the left torsion and right torsion modules in order to prove that it is artinian too. This duality occurs as a special case of a duality we find in a number of contexts, in particular, in the representation theory of finite dimensional algebras, so we shall develop it in some generality.

Let R be some ring, and let ${}_1\underline{B}$ be the full subcategory of f.p. modules of homological dimension 1, such that $\mathrm{Hom}_R(M,R) = 0$; we call these the <u>left bound modules</u>; similarly, we define the <u>right bound modules</u>, that are the objects of the category $\underline{B}_1$. Our conditions simply mean that if M is in ${}_1\underline{B}$ if and only if M is the cokernel of some map $\alpha : P \to Q$ such that α and $\alpha^x : Q^x \to P^x$ are injective; it is clear that this should lead to a duality between ${}_1\underline{B}$ and $\underline{B}_1$ by sending coker α to coker α^x.

<u>Theorem 1.20</u> The categories ${}_1\underline{B}$ and $\underline{B}_1$ are dual with respect to the functors $\mathrm{Ext}^1_R(\_,R)$.

Proof: Let $0 \to P \overset{\alpha}{\to} Q \to M \to 0$ be a presentation by f.g. projectives of an element of ${}_1\underline{B}$; then, $0 \to Q \overset{\alpha^x}{\to} P^x \to \mathrm{Ext}^1_R(M,R) \to 0$, is a presentation of $\mathrm{Ext}^1_R(M,R)$, which we shall call the <u>transpose</u> of M, and write as $\mathrm{Tr}M$; it is clear that $\mathrm{Tr}M$ lies in $\underline{B}_1$, and that
$\mathrm{Tr}(\mathrm{Tr}M) = \mathrm{Ext}^1_R(\mathrm{Tr}M,R) \cong \text{coker } \alpha \cong M$.

$\mathrm{Ext}^1_R(\_,R)$ is a contravariant functor of the first variable, so $\beta : M \to N$ induces a map $\mathrm{Tr}\,\beta\colon \mathrm{Tr}N \to \mathrm{Tr}M$; we write this out explicitly in order to show that the diagram below is commutative:

$$\begin{array}{ccc} \mathrm{Tr}(\mathrm{Tr}M) & \xrightarrow{\ \mathrm{Tr}(\mathrm{Tr}\beta)\ } & \mathrm{Tr}(\mathrm{Tr}N) \\ \wr\Vert & & \wr\Vert \\ M & \xrightarrow{\ \beta\ } & N \end{array}$$

, where the vertical maps are simply given by the isomorphisms coker $(\alpha^x)^x \cong$ coker α.

Let $0 \to P_1 \overset{\alpha_1}{\to} Q_1 \to M \to 0$, and $0 \to P_2 \overset{\alpha_2}{\to} Q_2 \to N \to 0$ be presentations of M and N; then we have the commutative diagram below:

$$\begin{array}{ccccccccc} 0 & \to & P_1 & \overset{\alpha_1}{\to} & Q_1 & \to & M & \to & 0 \\ & & \downarrow & & \downarrow & & \downarrow\beta & & \\ 0 & \to & P_2 & \overset{\alpha_2}{\to} & Q_2 & \to & N & \to & 0 \end{array}$$

Dualising gives us:

$$\begin{array}{ccccccccc}
0 & \to & Q_2^x & \xrightarrow{\alpha_1^x} & P_2^x & \to & \mathrm{TrN} & \to & 0 \\
 & & \downarrow & & \downarrow & & \downarrow{\scriptstyle \mathrm{Tr}\beta} & & \\
0 & \to & Q_1^x & \xrightarrow{\alpha_2^x} & P_1^x & \to & \mathrm{TrM} & \to & 0
\end{array}$$

which allows us to exhibit the map $\mathrm{Tr}\beta : \mathrm{TrN} \to \mathrm{TrM}$. Dualising again gives,

$$\begin{array}{ccccccccc}
0 & \to & P_1 & \xrightarrow{\alpha_1^{xx}} & Q_1 & \to & \mathrm{Tr(TrM)} & \to & 0 \\
 & & \downarrow & & \downarrow & & \downarrow{\scriptstyle \mathrm{TrTr}\beta} & & \\
0 & \to & P_2 & \xrightarrow{\alpha_2^{xx}} & Q_2 & \to & \mathrm{Tr(TrN)} & \to & 0 .
\end{array}$$

The natural equivalence on the category of f.g. projectives between the identity and the double dual induces a natural equivalence between the identity and the double transpose on $_1\underline{B}$ and on $\underline{B}_1$.

If R is a semihereditary ring and M is a f.g. module, $\mathrm{Hom}(M,R) = 0$ is equivalent to M's having no projective direct summand. If R is a semihereditary ring with a faithful rank function ρ and M is the cokernel of a full map with respect to ρ, then M lies in $_1\underline{B}$, since the dual of a full map is a full map, and full maps must be injective. It is clear that TrM is also the cokernel of a full map; consequently, Tr restricts to a duality between the categories of left and right torsion modules. In fact, it is clear that if we have a ring R with a faithful Sylvester rank function, we may define the category of left and right torsion modules and show that they are dual in the above way, since, under these circumstances, all full maps are injective.

We wish to show that these categories of torsion modules with respect to a projective rank function on a hereditary ring are both artinian and noetherian, and it is clear from the duality that we have found that we shall need to prove only one of these conditions, since the other is a direct consequence. So we shall need a certain type of ascending chain conditions on a ring.

<u>Lemma 1.21</u> Let R be a left $\aleph_0$-hereditary ring with a faithful projective rank function ρ. Let $P_0 \subseteq P_1 \subseteq P_2 \subseteq \ldots\ldots \subseteq Q$ be an ascending chain of

f.g. submodules of the left projective module Q such that $P_i \subseteq P_{i+1}$ is a full map with respect to ρ for all i; then the chain must stop after finitely many steps.

Proof: Consider $\bigcup_i P_i$ which is countably generated and so projective. Therefore, by 1.2, $\bigcup_i P_i \cong \bigoplus_j Q_j$, where each Q_j is f.g. projective.

Eventually, $P_0 \subseteq \bigoplus_{j=1}^{m} Q_j \subseteq P_n$, for some m and n. $\bigoplus_{j=1}^{m} Q_j$ is a direct summand of $\bigcup_i P_i$, and so, of P_n; hence $\rho(\bigoplus_{j=1}^{m} Q_j) \leq \rho(P_n)$ with equality only when $\bigoplus_{j=1}^{m} Q_j = P_n$, since the rank function is faithful. However, $P_0 \subseteq P_n$ is a full map, and equality must hold. Since this is true for all P_m, for $m' > n$, the chain must stop.

If R is a ring with a faithful projective rank function ρ and $\alpha : P \to Q$ is a full map with respect to ρ, we define it to be an atomic full map if, in any non-trivial factorisation,

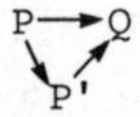

,

$\rho(P') > \rho(P) = \rho(Q) = \rho(\alpha)$.

We shall show that, over an $\aleph_0$-hereditary ring, any full map has a finite factorisation as a product of atomic full maps, that any two such factorisations have the same length, and the atomic full maps in one factorisation may be paired off with those in the other so that the corresponding maps are stably associated, which, as we saw earlier, is the same as their cokernels being isomorphic.

<u>Theorem 1.22</u> Let R be an $\aleph_0$-hereditary ring with a faithful projective rank function ρ. The category of left or right torsion modules with respect to ρ is an artinian and noetherian abelian category.

Proof: By theorem 1.18, it is an abelian category and, by theorem 1.22, it is a noetherian category; it is dual to the category of torsion modules on the other side by 1.19 and the subsequence discussion, so it must be artinian too.

<u>Theorem 1.23</u> Let R be an $\aleph_0$-hereditary ring with a faithful projective

rank function ρ. Then the atomic full maps with respect to ρ are exactly the maps whose cokernels are simple objects in the category of torsion modules. Consequently, every map has a factorisation into atoms; if $\alpha = \alpha_1\alpha_2 \dots \alpha_n$, and $\alpha = \beta_1\beta_2 \dots \beta_m$, where α_i, β_j are atoms, then $m = n$, and there is a permutation in S_n such that α_i is stably associated to $\beta_{\sigma(i)}$.

Proof: It is clear that only the cokernels of atoms can be simple in the category of torsion modules with respect to ρ, and that the cokernels of atoms are simple.

Every object in an artinian and noetherian abelian category has a finite composition series consisting of simple modules, any two such series have the same length and the simple modules may be paired off between the two series so that two paired together are isomorphic: remembering that if the full map α_i has the same cokernel as β_j, that they must be stably associated, and that a composition series for the cokernel of a full map in the category of torsion modules corresponds to a factorisation of the map as a product of simples, we deduce the theorem.

It follows at once from this that if α is an atomic full map, then the ring of endomorphisms of coker α is a skew field, since it is a simple object in an abelian category.

2 THE COPRODUCT THEOREMS

The purpose of this chapter is to present statements of Bergman's coproduct theorems (Bergman 74) and then to use these results as Bergman does in (Bergman 74') to study a number of interesting ring constructions. Both of these papers by Bergman are of great importance and this chapter is not designed to obviate the need to read them, but rather to interest the reader in their results and to make them plausible; however, we shall summarise all that we need from them for the majority of this book. At the end of the chapter, there is a discussion of the commutative analogues of some of the constructions considered.

The basic coproduct theorems

For ease of notation, we shall consider right modules in this chapter. We begin by running through a number of definitions that we need in order to state the coproduct theorems.

An R_0-ring is a ring R with a specified homomorphism from R_0 to R; it is a faithful R_0-ring if this is an embedding. Given a family of R_0-rings $\{R_\lambda : \lambda \in \Lambda\}$, there is a coproduct in the category of R_0-rings, which we shall write as $R = \sqcup_{R_0} R_\lambda$, and call the ring coproduct of the family of rings $\{R_\lambda : \lambda \in \Lambda\}$ amalgamating R_0. There is no reason to suppose in the general case that this ring is not the trivial ring. We call R_0 the base ring of the coproduct and the rings R_λ are the factor rings of the coproduct R. It is technically convenient to label $\Lambda \cup \{0\}$ as M, and to use μ for a general element of M. Most of the time, we shall take R_0 to be a semisimple artinian ring, and each R_λ will be a faithful R_0-ring; we shall see that under these conditions each R_λ embeds in the coproduct R.

Ideally, we should like to be able to reduce all problems about the module theory of the coproduct to the module theory of the factor rings; in the case where R_0 is semisimple artinian and each R_λ is a faithful R_0

ring we can go a long way towards doing this; in fact, most practical problems about such coproducts may be solved although the proof is often rather technical.

An induced module over the coproduct R is a module of the form $\oplus_\mu M_\mu \otimes_{R_\mu} R$ where M_μ is a right R_μ module; an induced module of the form $M_0 \otimes_{R_0} R$ or an R_λ module of the form $M_0 \otimes_{R_0} R_\lambda$ is called a basic module. We say that a map of the form $\oplus_\mu \alpha_\mu \otimes_{R_\mu} R : \oplus_\mu M_\mu \otimes_{R_\mu} R \to \oplus_\mu N_\mu \otimes_{R_\mu} R$ is an induced map.

There are certain induced modules which are clearly isomorphic over R; let $M = \oplus_\mu M_\mu \otimes_{R_\mu} R$ be an induced module and suppose that for some λ_1, we have an isomorphism $M_{\lambda_1} \cong M'_{\lambda_1} \oplus (M_0 \otimes_{R_0} R_{\lambda_1})$; then $M \cong N$, where $N = \oplus_\mu N_\mu \otimes_{R_\mu} R$, where $M_\lambda \cong N_\lambda$ for $\lambda \neq \lambda_1, \lambda_2$; $N_{\lambda_1} \cong M'_{\lambda_1}$; and $N_{\lambda_2} \cong M_{\lambda_2} \oplus (M_0 \otimes_{R_0} R_{\lambda_2})$. Such an isomorphism of induced modules is called a basic transfer map; if R_0 is a skew field it is usually called a free transfer map.

There is one further type of isomorphism that we shall need to consider, a particular sort of automorphism of an induced module. Let $M = \oplus_\mu M_\mu \otimes_{R_\mu} R$, and assume that for some μ_1 there is a linear functional $e : M_{\mu_1} \to R_{\mu_1}$. Extend e to a linear functional $e : M \to R$ by setting $e(M_\mu) = 0$ for $\mu \neq \mu_1$, and then extending by linearity. Let a be in R, and write $l_a : R \to R$ for left multiplication by a. Finally, for some μ_2, let x be in M_{μ_2}, and let $\gamma : R \to M$ take 1 to x. Then, if $\mu_1 \neq \mu_2$, the composite $el_a\gamma$ has square 0, and if $\mu_1 = \mu_2$ we ensure this by specifying that x should lie in the kernel of e. The map $I_M - el_a\gamma$ is invertible; we call such an automorphism a transvection.

We may now state the main theorems. We shall assume that R_0 is a semisimple artinian ring, and that each R_λ is a faithful R_0-ring, for the rest of the chapter.

Theorem 2.1 Let R_0 be a semisimple artinian ring, and $\{R_\lambda : \lambda \in \Lambda\}$, a family of faithful R_0-rings. Let $R = \sqcup_{R_0} R$. If $M = \oplus_\mu M_\mu \otimes_{R_\mu} R$ is an induced module, each M_μ embeds in M, and M is isomorphic as an R_μ-module to a direct sum of M_μ and a basic module.

So we can write $R_\mu \subseteq R$, and $M_\mu \subseteq M$.

Theorem 2.2 Let R_0 be semisimple artinian, and $\{R_\lambda : \lambda \in \Lambda\}$ a family of faithful R_0-rings. Let $R = \sqcup_{R_0} R_\lambda$. Then any R-submodule of an induced module is isomorphic to an induced module.

Theorem 2.3 Let $R = \sqcup_{R_0} R_\lambda$, where R_0 is semisimple artinian, and each R_λ is a faithful R_0-ring. Let $f : M \to N$ be a surjection of induced modules; then there is an isomorphism of induced modules $g : M' \cong M$, which is a finite composition of basic transfers, and transvections such that the composite $gf : M' \to N$ is an induced map.

These three theorems allow us to deduce all the rest of those results that we shall prove in this chapter; however, in later work, we shall need technical versions of these theorems, and in order to state these results, we shall need to set up a certain amount of the machinery to prove the co-product theorems. We shall do this under the assumption that R_0 is a skew field, and indicate the modifications needed to deal with the general case.

We assume that R_0 is a skew field; then each R_μ is a free right R_0 module, and we can choose a basis for R_λ of the form $\{1\} \cup T$. Write $T = \bigcup_\lambda T_\lambda$. For each N_λ in the induced module $N = \bigoplus_\mu N_\mu \otimes_{R_\mu} R$ we pick an R_0 basis, S_μ. Write $S = \bigcup_\mu S_\mu$. If t is in T_λ, we say that it is associated to λ; if s is in S_λ, we say that it is associated to λ; if s is in S_0, we say that it is associated to no index. A monomial is a formal product $st_1t_2\ldots t_n \ldots s \in S$, $t_i \in T_1$ or else an element of S_1 such that no two successive terms in the series $s, t_1, \ldots t_n$, are associated to the same index. Let U be the set of monomials; an element of U is associated to λ, if and only if its last factor is associated to λ. Every element of U is associated to some index except for those in S_0. We denote by $U_{\sim\lambda}$, those elements of U that are not associated to λ.

Theorem 2.4 (see theorem 2.1) Let R_0 be a skew field, and let $\{R_\lambda : \lambda \in \Lambda\}$ be a family of R_0-rings. Let $R = \sqcup_{R_0} R_\lambda$; then the induced module $N = \bigoplus_\mu N_\mu \otimes_{R_\mu} R$ has for a right R_0-basis the set U, defined in the foregoing material. For each λ, N is the direct sum as right R_λ-module of N_λ and a free right R_λ-module on the basis $U_{\sim\lambda}$.

Given λ and $u \in U_{\sim\lambda}$, we denote by $c_{\lambda u} : N \to R_\lambda$, the R_λ-linear right 'co-efficient of u' map given by the decomposition of N in the last theorem; for $u \in U$, we denote by $c_{Ou} : N \to R_O$, the R_O-linear 'right coefficient of u' map given by the decomposition of N as an R_O-module in the last theorem. For $\lambda \in \Lambda$, the λ-support of an element x in N relative to the decomposition of N in the last theorem, is the set of elements $u \in U_{\sim\lambda}$ such that $c_{\lambda u}(x)$ is not 0; x has empty λ-support if and only if it lies in N_λ. The 0-support (or support) of an element consists of those element $u \in U$ such that $c_{Ou}(x)$ is not 0.

The degree of a monomial $st_1 \ldots t_n$ is defined to be $(n+1)$, and the degree of an element of S is 1; the degree of an element of N is the maximum degree of a monomial in its support. We define an element x in N to be λ-pure, if all those monomials in its support of maximal degree are associated to λ. It is 0-pure if it is not λ-pure for any λ.

Next, we well-order the sets S and T in some manner; this induces a well-ordering on U, first by degree and then lexicographically, reading from left to right.

The leading term of an element in N (with respect to the fixed ordering) is the maximal element in its support. We call this the 0-leading term. If some element x is not λ-pure, then some elements in its support of maximal degree will not be associated to λ; we define the λ-leading term of x to be the maximal element of this sort.

<u>Theorem 2.5</u> (see theorem 2.2) Let R_O be a skew field and let $R = \underset{R_O}{\sqcup} R_\lambda$. Let N be the induced module $\underset{\mu}{\oplus} N_\mu \otimes_{R_\mu} R$, and let S,T and U be defined as in the foregoing discussion. Let L be some R-submodule of N. Define L_μ to be the R_μ-submodule of L, consisting of those elements whose μ-support does not contain the μ-leading term of some non-μ-pure element of L. Then $L \cong \underset{\mu}{\oplus} L_\mu \otimes_{R_\mu} R$, in the natural way.

We shall outline the adjustments needed to deal with the case that R_O is a semisimple artinian ring. The first step is to use Morita equivalence to pass from this case to the particular situation that R_O is a direct sum of skew fields. This technique is of interest and use in its own right.

Let R_O be a semisimple artinian, and let $S_1, S_2, \ldots S_n$ be a complete set of simple R_O modules such that $S_i \cong S_j$ $i = j$. The projective module $P = \underset{i}{\oplus} S_i$ is a projective generator over R_O; so by Morita equivalence the category of modules over R_O is naturally equivalent to the category of modules over $\bar{R}_0 = \mathrm{End}_{R_0}(P)$. Since each S_i is simple and they are mutually non-isomorphic $\bar{R}_0$ is the direct sum of the endomorphism rings of S_i, and so, it is a direct sum of skew fields.

Let R_λ be a faithful R_0-ring; then $P \otimes_{R_0} R_\lambda$ is a projective generator over R_λ, so that the category of modules over R_λ is naturally equivalent to the category of modules over $\bar{R}_\lambda = \mathrm{End}_{R_\lambda}(P \otimes_{R_0} R_\lambda)$, and there is a natural embedding of $\bar{R}_0$ in $\bar{R}_\lambda$.

We give an example to clarify what is happening here. Let $R_0 = M_2(k) \times k$; in this case $\bar{R}_0$ is simply $k \times k$. Let $R_1 = M_3(k)$, where R_1 is an R_0-ring via

$$\left(\begin{pmatrix} a & b \\ c & d \end{pmatrix}, e\right) \to \begin{pmatrix} a & b & 0 \\ c & d & 0 \\ 0 & 0 & e \end{pmatrix}$$

Then, $\bar{R}_1 = M_2(k)$, and it is an $\bar{R}_0$-ring via $(a,b) \to \begin{pmatrix} a & 0 \\ 0 & b \end{pmatrix}$

Returning to the general case, we are given a family of faithful R_0-rings, $\{R_\lambda : \lambda \in \Lambda\}$ and we have associated to this a family of faithful $\bar{R}_0$-rings, $\{\bar{R}_\lambda : \lambda \in \Lambda\}$. We wish to study the ring $R = \coprod_{R_0} R_\lambda$; and, as before, we form the $\bar{R}_0$-ring, $\bar{R} = \mathrm{End}_R(P \otimes_{R_0} R)$; it is clear that this is simply the ring $\coprod_{R_0} \bar{R}_\lambda$.

Therefore, in order to study the category of R modules, we may as well study the category of $\bar{R}$ modules, which is a coproduct over a direct sum of skew fields. All the statements of the next two theorems are Morita invariant and so they translate well.

So for the present, we assume that R_0 is a direct sum of skew fields, $R_0 \cong \prod_{i=1}^{n} K_i$; Let $\{e^i : i = 1 \to n\}$ be the complete set of orthogonal central idempotents. Any R_0 module has a decomposition as a direct sum of vector spaces over K^i, $M = \bigoplus_{i=1}^{n} Me^i$; we choose a basis B^i for each Me^i as a vector space over K^i. We call the n-tuple of bases $\{B^i\}$ a basis for M over R_0.

If R_λ is an R_0-ring, it decomposes as a right R_λ-module, $R_\lambda \cong \bigoplus_{i=1}^{n} e^i R_\lambda$; we write ${}^iR_\lambda = e^i R_\lambda$. In turn, ${}^iR$ decomposes as a right R_0-module as ${}^iR \cong \bigoplus_{i=1}^{n} {}^iRe^j$; again, we write ${}^iR_\lambda{}^j = e^i R_\lambda e^j$, we note that the usual formalism of matrix units works with respect to these superscripts. We pick a basis for ${}^{i}R_{\lambda}^{j}$ as right K^j module, ${}^{i}T_{\lambda}^{j}$ for $i \neq j$, and ${}^{i}T^{i} \cup \{e^i\}$ for the remaining cases. We note that this gives us a basis for R_λ as R_O-module in the sense defined earlier.

Let $\{N_\mu : \mu \in M\}$ be a family of R_μ-modules. We choose an R_O-basis for each $N_\mu, \{S_\mu^i\}$. Let $S = \bigcup_{i,\mu} S_\mu^i$ and $T = \bigcup_{\lambda,i,j} {}^{i}T_{\lambda}^{j}$. For each t in ${}^{i}T_{\lambda}^{j}$, i and j are respectively the right and the left index of t; λ is the λ-index of t. A member of S^i has the right index i, and if it does not lie in N_O, it has the natural λ-index.

Let U be S and the set of formal products $st_1, \ldots t_n$, where $s \in S$, $t_i \in T$, where adjacent terms do not have the same λ-index, but the right index of any term is equal to the left index of the next term. The right index or λ-index of any such element is the right or λ-index of its last term. So we may partition $U = \bigcup_i U^i$, where U^i is the set of elements associated to i; we form the free K^i module N^i on U^i and consider the R_O module $N = \bigoplus_i N^i$.

Let $U_{\sim\lambda}^i$ be those elements of U^i not associated with λ. We may form the R_λ module $N_\lambda \oplus (\bigoplus_{i=1}^n U^i K^i \otimes_{R_O} R_\lambda)$.

<u>Theorem 2.6</u> Let R_O be a direct sum of skew fields, $\times_{i=1}^n K_i$. Let $\{\mathbf{R}_\lambda : \lambda \in \Lambda\}$ be a family of faithful R_O-rings, and let $\{N_\mu : \mu \in M\}$ be a collection of R_μ-modules. Then $\bigoplus_\mu N_\mu \otimes_{R_\mu} R$, where $R = \sqcup_{R_O} R_\lambda$, is isomorphic to N as defined above as R_O-module, and as R_λ module it is isomorphic to $N_\lambda \oplus (\oplus U^i K^i \otimes_{R_O} R_\lambda)$, where U^i in this last representation is identified in N with the same subset of U.

We note that for $u \in U^i$, uR is isomorphic to $e^i R$ via a map sending u to e^i. Consequently, the definition given in 2.6 allows us to define co-ordinate functions $c_{\mu u} : N \to R_\mu$, which takes values in $e^i R$, where i is the right index of u.

As we did when R_O was simply a skew field, we well-order S and T, and then well-order U length lexicographically. We define degree, μ-purity, the μ-leading term and so on as we did previously. To recover a version of coproduct theorem 2.5, we need one more concept, that of homogeneity. Given an R_O module M, we define $m \in M$ to be i-<u>homogeneous</u> if $me^i = m$. An element is <u>homogeneous</u> if it is i-homogeneous for some i.

Theorem 2.7 Let R_0 be a direct sum of skew fields $\overset{n}{\underset{i=1}{\times}} K^i$; let $\{R_\lambda : \lambda \in \Lambda\}$ be a family of faithful R_0-rings, and let $R = \sqcup_{R_0} R_\lambda$. Let N be the induced module $\oplus_\mu N_\mu \otimes_{R_\mu} R$, and let L be an R-submodule of N. Let L_μ^i be the R_0-submodule of L consisting of those i-homogeneous elements of L, whose μ-support does not contain the μ-leading term of some homogeneous non-μ-pure element of L; then $\overset{n}{\underset{i=1}{\oplus}} L_\mu^i$ is an R_μ submodule L_μ of L, and $L \cong \oplus_\mu L_\mu \otimes_{R_\mu} R$.

In the more general case, where R_0 is a semisimple artinian ring, each R is a faithful R_0-ring, $R = \sqcup_{R_0} R_\lambda$, and $N = \oplus_\mu N_\mu \otimes_{R_\mu} R$ with an R-submodule L, we pass by Morita equivalence denoted by bars to the case described by the hypotheses of the last theorem. We then find distinguished submodules $\bar{L}_\mu$ such that $\bar{L} \cong \oplus_\mu \bar{L}_\mu \otimes_{\bar{R}_\mu} \bar{R}$; by Morita equivalence, we have submodules L_μ such that $L \cong \oplus_\mu L_\mu \otimes_{R_\mu} R$.

We note a small corollary of 2.7, which will have some consequence later.

Lemma 2.8 Let $R = \sqcup_{R_0} R_\lambda$, where R_0 is semisimple artinian, and each R_λ is a faithful R_0-ring. Let L be an R-submodule of the induced module $\oplus_\mu N_\mu \otimes_{R_\mu} R$; then, in the decomposition of L given in 2.7, and the following discussion, $L \cong \oplus_\mu L_\mu \otimes_{R_\mu} R$, we find that $L \cap N_\mu \subseteq L_\mu$.

Proof: By the process of Morita equivalence, it is enough to show this, when R_0 is a direct sum of skew fields. In this case, it is clear, for, if $l \in L \cap N_\mu$, le^i is i-homogeneous and has empty μ-support so it satisfies the conditions to be in L_μ.

There is another result of this technical type that we shall need later on.

Lemma 2.9 Let $R = \sqcup_{R_0} R_\lambda$, where R_0 is a skew field. Let L be an R submodule of an induced module $N = \oplus_\mu N_\mu \otimes_{R_\mu} R$; write $L = \oplus L_\mu \otimes_{R_\mu} R$ given by theorem 2.5, and assume that L_0 is empty. Then if $\Sigma n_\lambda \in L$ for $n_\lambda \in N_\lambda$, each n_λ must be in L.

Proof: We may assume that at least two of the n_λ are non-zero; so Σn_λ is 0-pure. Write the support of Σn_λ with respect to our well-ordered basis in descending order. Given two finite strings of basis elements in descending order $\{a_i\}$ and $\{b_i\}$ we shall say that $\{b_i\}$ is less than $\{a_i\}$ if for the first place at which they differ $b_i < a_i$ or else $\{b_i\}$ is an initial string of $\{a_i\}$. We choose Σn_λ in L so that $n_\lambda \in L$ and the associated string of basis elements is minimal in the above sense. Σn_λ is 0-pure but $\Sigma n_\lambda \notin L_0$; so, it contains the leading term of some pure element which is forced to take the form n'_λ for some $n'_\lambda \in N_\lambda$; by subtracting an appropriate multiple of n'_λ from Σn_λ we reduce the support and obtain a contradiction.

It is clear that the information we have built up in the preceding results allows us to answer a number of natural questions about ring coproducts, amalgamating a common semisimple artinian ring. We begin with a few such applications.

<u>Theorem 2.10</u> Let $R = \sqcup_{R_0} R_\lambda$, where R_0 is semisimple artinian and each R_λ is a faithful R_0-ring. Then the homological (or weak) dimension of $M \otimes_{R_\lambda} R$ over R is equal to the homological (or weak) dimension of M over R_λ.

Proof: Given a resolution $\underline{P} \to M \to 0$ over R_λ, $\underline{P} \otimes_{R_\lambda} R \to M \otimes_{R_\lambda} R \to 0$ is a resolution of $M \otimes_{R_\lambda} R$ over R, since by 2.1, R is flat over R_λ. Again, by 2.1, this resolution of $M \otimes_{R_\lambda} R$ considered over R_λ contains $\underline{P} \to M \to 0$ as a direct summand, so that the homological dimension of M cannot decrease on passing to $M \otimes_{R_\lambda} R$.

From this, we are able to determine the global and weak dimensions of a ring coproduct.

<u>Theorem 2.11</u> Let $R = \sqcup_{R_0} R_\lambda$, where R_0 is semisimple artinian, and each R_λ is a faithful R_0-ring; then the right global dimension of R is equal to the supremum of the right global dimensions of the R_λ, provided that one of them is not 0. If each R_λ is semisimple, however, the global dimension may be 0 or 1. A similar result holds for weak dimension.

Proof: Consider any submodule of a free R module; by 2.2, it is an induced module. Moreover, by 2.1, if $M \cong \bigoplus_{\mu} M_{\mu} \otimes_{R_\mu} R$. each M_{μ} is a submodule of a basic module, which are projective modules. From theorem 2.9, it follows that the global dimension of R is the supremum of the global dimensions of the R_μ, provided that one of them is not semisimple artinian. If each of them is semisimple artinian, it follows that the global dimension is at most 1. Exactly the same proof holds for the weak dimension.

If R_0 is a skew field, it is fairly clear that R cannot be a semisimple artinian ring; however, this is most easily shown after we have developed some more properties of the ring coproduct, so we shall deal with this towards the end of the chapter. However, we present an example to show that when R_0 is only semisimple artinian, the ring coproduct may be simple artinian.

Let $R_0 = k \times k \times k$, $R_1 = M_2(k) \times k$, and $R_2 = k \times M_2(k)$, where

$$R_0 \to R \quad \text{by} \quad (a,b,c) \quad \left(\begin{pmatrix} a & 0 \\ 0 & b \end{pmatrix}, c\right), \text{ and}$$

$$R_0 \to R_2 \quad \text{by} \quad (a,b,c) \quad \left(a, \begin{pmatrix} b & 0 \\ 0 & c \end{pmatrix}\right)$$

It is clear that $R_1 \underset{R_0}{\sqcup} R_2 \cong M_3(k)$.

One can show by the same method that the coproduct of right semihereditary rings is right semihereditary and a similar result holds for χ_0-hereditary.

Before passing from the study of the category of modules over the coproduct to the more particular study of the category of f.g. projectives, we leave as an exercise for the reader the details of the next example, which answers a question of Bergman.

Example 2.12 We wish to find an extension of induced modules that is not induced.

Consider the free ring $R = k\langle x,y \rangle \cong k[x] \underset{k}{\sqcup} k[y]$. The module R/xyR is an extension of the induced module R/yR by the induced module R/xR, but it is not an induced module itself.

We start to look at the category of f.g. projectives over the ring coproduct. We already know all the objects of this category. We have the next result.

Theorem 2.13 Let $R = \sqcup_{R_0} R_\lambda$, where R_0 is semisimple artinian, and each R_λ is a faithful R_0-ring. Then, a projective module over R has the $\oplus P_\mu \otimes_{R_\mu} R$, where each P_μ is a projective module over R_μ. The monoid of isomorphism classes of f.g. projectives over R is isomorphic to the commutative monoid coproduct of the monoids of isomorphism classes of f.g. projectives over R_λ, amalgamating the monoid of isomorphism classes of f.g. projectives over R_0. Consequently, $K_0(R)$ is the commutative group coproduct of $K_0(R_\lambda)$, amalgamating $K_0(R_0)$.

Proof: Every projective module lies in a free module and so must be isomorphic to an induced module by theorem 2.2. By theorem 2.9, we know that if $P \cong \oplus_\mu M_\mu \otimes_{R_\mu} R$, each M_μ must be a projective R_μ module.

Theorem 2.3 tells us that any isomorphism between two f.g. induced modules is the composite of a finite sequence of basic transfers and transvections followed by an induced map which, over R, is an isomorphism and so must have been an isomorphism on each summand. Transvections do not alter the form of an induced module; basic transfers simply amount to the amalgamation of the images of $P_\oplus(R_0)$ in the different $P_\oplus(R_\lambda)$, which proves the result for the monoids of f.g. projectives. The result for K_0 follows since K_0 is the universal abelian group functor applied to $P_\oplus$.

We should also like to have information on the maps between f.g. projectives; that is, we should like to be able to explain all isomorphisms and all zero-divisors over the coproduct in terms of the factor rings. We already know how to explain all isomorphisms since 2.3 explains all surjections, so we look at the zero-divisors next.

Theorem 2.14 Let $R = \sqcup_{R_0} R$, where R_0 is a semisimple artinian ring and each R_λ is a faithful R_0-ring. Let $\alpha : P \to P'$, and $\beta : P' \to P''$ be maps between f.g. projectives over R such that $\alpha\beta = 0$. Then there exists a commutative diagram:

$$\begin{array}{ccccc}
P & \xrightarrow{\alpha} & P' & \xrightarrow{\beta} & P'' \\
\downarrow & & \cong & & \cup \\
\underset{\mu}{\oplus} P_\mu \otimes_{R_\mu} R & \xrightarrow{\oplus \alpha_\mu \otimes R} & \underset{\mu}{\oplus} P'_\mu \otimes_{R_\mu} R & \xrightarrow{\oplus \beta_\mu \otimes R} & \underset{\mu}{\oplus} P''_\mu \otimes_{R_\mu} R
\end{array}$$

where each direct sum in this diagram is a finite direct sum of f.g. projectives and $\alpha_\mu \beta_\mu = 0$, for all μ.

Proof: First, $\mathrm{im}\beta$ is an induced module by 2.2; so $P' \to \mathrm{im}\beta$ is a surjection of induced modules, and by 2.3, there exists a commutative diagram:

$$\begin{array}{ccccc}
P & \xrightarrow{\alpha} & P' & \xrightarrow{\beta} & P'' \\
\downarrow{\scriptstyle \alpha'} & & \cong & & \cup \\
\underset{\mu}{\oplus} \ker\beta_\mu \otimes R & \longrightarrow & \underset{\mu}{\oplus} P'_\mu \otimes_{R_\mu} R & \longrightarrow & \mathrm{im}\beta_\mu \otimes_{R_\mu} R \ = \ \mathrm{im}\beta
\end{array}$$

So the problem is to replace $\ker\beta_\mu$, and $\mathrm{im}\beta_\mu$ by suitable f.g projectives. $\mathrm{im}\alpha'$ is a f.g. submodule of $\underset{\mu}{\oplus} \ker\beta_\mu \otimes R$ and so lies in an R-submodule generated by finitely many elements from $\ker\beta_\mu$. Thus, there is an induced map from $\underset{\mu}{\oplus} P_\mu \otimes_{R_\mu} R$ to $\underset{\mu}{\oplus} \ker\beta_\mu \otimes R$, where each P_μ is a finitely generated free R_μ-module, such that the image of this induced map contains the image of α'. Since P is a projective module, we have a commutative diagram:

$$\begin{array}{ccccc}
P & \longrightarrow & P' & \longrightarrow & P'' \\
\downarrow & & \cong & & \cup \\
\underset{\mu}{\oplus} P_\mu \otimes_{R_\mu} R & \longrightarrow & \underset{\mu}{\oplus} P'_\mu \otimes_{R_\mu} R & \longrightarrow & \mathrm{im}\beta_\mu \otimes_{R_\mu} R
\end{array}$$

We miss out the bottom right hand corner of this diagram ($\underset{\mu}{\oplus} \mathrm{im}\beta_\mu \otimes_{R_\mu} R$); dualise the rest of the diagram and then fill in the bottom left hand corner of this diagram as we did in order to replace $\underset{\mu}{\oplus} \ker\beta_\mu \otimes_{R_\mu} R$. The diagram we obtain is the dual of a diagram that suits our needs.

Universal ring constructions

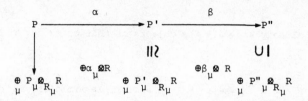

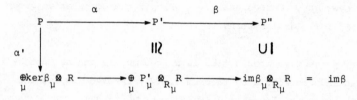

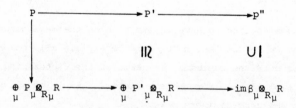

These results on the coproduct construction allow us to study a number of interesting constructions on the category of f.g. projectives over a ring. Given a k-algebra, S, the ring coproduct $S \underset{k}{\sqcup} k[x]$ may be regarded as the S-ring with a universal map on the free module of rank 1 that centralises the k-algebra structure. Suppose that P_1 and P_2 are f.g. projectives over the ring S; we are also interested in finding and studying an S-ring, T, with a universal map from $P_1 \otimes_S T$ to $P_2 \otimes_S T$ that centralises the k-structure. In the language of functors, we are trying to find the S-ring in the category of k-algebras that represents the functor $S' \to \mathrm{Hom}_{S'}(P_1 \otimes_S S', P_2 \otimes_S S')$. We say that an object $\underline{O}$ in the category $\underline{C}$ represents a covariant functor $F : \underline{C} \to \underline{\text{Sets}}$ if $F(\ )$ is naturally equivalent to $\mathrm{Hom}_{\underline{C}}(\underline{O},\ )$. We should like to investigate other universal constructions; thus, we wish to find an S-ring, T_1, in the category of k-algebras with a universal isomorphism between $P_1 \otimes_S T_1$ and $P_2 \otimes_S T_1$. That is, T_1 represents the functor that associates to each S-ring that is a k-algebra the set of isomorphisms between $P_1 \otimes_S S'$ and $P_2 \otimes_S S'$. Again, we should like to find an S-ring, T_2, that is a k-algebra with a universal idempotent map on the projective $P \otimes_S T_2$; that is, T_2 represent the functor on the category of S-rings that are k-algebras, which associates to each object S' the set of idempotent maps on $P \otimes_S S'$.

It is quite easy to see that there are rings with these universal properties, essentially by generator and relation constructions. First, we note that if $\underline{A}$ is an additive category such that every object is a direct summand of F^n for some object F in the category, then this category is a full subcategory of the category of f.g. projectives over the endomorphism ring of F; F becomes the free module of rank 1. So, if we start off with a ring R, and adjoin some set of maps to the category of f.g. projectives over R subject to some set of relations, and construct the additive category that they generate, the objects are still all direct summands of R^n for suitable n, and therefore, the category is a full subcategory of the category of f.g. projectives over the endomorphism ring of the object R^1. Clearly, there is a homomorphism from R to this endomorphism ring E, which makes E into the universal R-ring having this additional set of maps satisfying the specified set of relations. Clearly, the constructions in the preceding paragraph all have this form, and we shall discuss in chapter 4 another construction which may also be shown to exist by this method. For our purposes in this chapter it is more useful to study them in a different way.

One construction of a slightly different nature that we wish to study is the formation of the universal k-algebra T' with a universal homomorphism from S to $M_n(T')$. Here, T' represents the functor on the category of k-algebras, $\mathrm{Hom}_{k\text{-alg}}(S, M_n(\ ))$.

The constructions on the category of f.g. projectives are all approached in a similar way. We shall explain the case of adjoining a universal map between two projectives in some detail, and since the other cases use similar arguments, we shall consider them more briefly.

Let $\phi : R \to S$ be a homomorphism between k-algebras, and let Q_1 and Q_2 be f.g. projectives; then, if R' is the universal R-ring, and k-algebra, with a universal map from Q_1 to Q_2, it is clear that $R' \sqcup_R S$ is the universal S-ring and k-algebra, with a universal map from $Q_1 \otimes_R S$ to $Q_2 \otimes_R S$.

For suitably large n, there are orthogonal idempotents, e_1, e_2 in $M_n(S)$, such that $e_i({}^nS) \cong P_i$; under the Morita equivalence of S and $M_n(S)$, $e_i({}^nS)$ becomes $e_iM_n(S)$, so it is enough to show that there is an $M_n(S)$-ring with a universal map from $e_1M_n(S')$ to $e_2M_n(S')$ for an S-ring, S', that is a k-algebra. Let e_3 be that idempotent such that $e_1 + e_2 + e_3 = 1$ in $M_n(S)$; we have a map $k \times k \times k \to M_n(S)$ given by $(a,b,c) \to e_1a + e_2b + e_3c$, which makes $M_n(S)$ into a faithful $k \times k \times k$-ring. The projective module induced by the first summand of $k \times k \times k$ is $e_1M_n(S)$ and that induced by the second summand is $e_2M_n(S)$. It is easily seen that the $k \times k$-ring with a universal map from $k \times O$ to $O \times k$ is the lower triangular matrix ring $T_2(k) = \begin{pmatrix} k & O \\ k & k \end{pmatrix}$, where $k \times k$ embeds along the diagonal, and e_{21} induced the universal map. Therefore, the ring coproduct, $T_2(k) \times k \underset{k \times k \times k}{\sqcup} M_n(S)$ is the universal $M_n(S)$-ring that is a k-algebra with a map from $e_1M_n(S)$ to $e_2M_n(S)$. By Morita equivalence, we have a universal S-ring with a universal map from P_1 to P_2. We shall use the symbol $T = S_k\langle\alpha:P_1 \to P_2\rangle$ for this construction. We notice that it is made for studying in terms of the ring coproduct.

<u>Theorem 2.15</u> Let S be a k-algebra and let P_1 and P_2 be f.g. projectives over S. Let $T = S_k\langle\alpha:P_1 \to P_2\rangle$ be the k-algebra and S-ring with a universal map from $P_1 \otimes_S T$ to $P_2 \otimes_S T$; then T has the same global or weak dimension as S except when the global or weak dimension of S is O; in this case, T has dimension 1. All f.g. projectives are induced from S; in fact, the ring homomorphism from S to T induces an isomorphism $P_\oplus(S) \cong P_\oplus(T)$.

Proof: We retain the notation of the preceding discussion.

We see that $M_n(T) \cong M_n(S) \underset{k\times k\times k}{\sqcup} T_2(k)\times k$; $T_2(k)$ has global and weak dimension 1, so that $M_n(T)$ has global or weak dimension equal to that of S except when S has dimension 0, in which case, T has dimension 1 by 2.10. The monoid of isomorphism classes of f.g. projectives over $T_2(k)\times k$ is isomorphic to the monoid of isomorphism classes of f.g. projectives over $k\times k\times k$, simply by the map induced by the ring homomorphism described earlier; so, from 2.12, we deduce that $P_{\oplus}(M_n(S) \underset{k\times k\times k}{\sqcup} T_2(k)\times k) \cong P_{\oplus}(M_n(S)$, from which it follows by Morita equivalence, that $P_{\oplus}(T) \cong P_{\oplus}(S)$.

The $k\times k$-ring with a universal isomorphism between $k\times 0$ and $0\times k$, is $M_2(k)$, where $k\times k$ embeds as the diagonal matrices; the elements e_{21} and e_{12} are the universal map and its inverse. Therefore, if e_1 and e_2 are orthogonal idempotents in $M_n(S)$ such that $e_i{}^nS \cong P_i$, the $M_n(S)$-ring with a universal isomorphism between $e_1M_n(S)$ and $e_2M_n(S)$ is just $M_n(S) \underset{k\times k\times k}{\sqcup} M_2(k)\times k$, where the map $k\times k\times k$ to $M_n(S)$ is $(a,b,c) \to ae_1 + be_2 + ce_3$, where e_3 is $1 - e_1 - e_2$, and $k\times k\times k$ to $M_2(k)\times k$ is given by $(a,b,c) \to \left(\begin{pmatrix} a & 0 \\ 0 & b\end{pmatrix}, c\right)$. By Morita equivalence, we have an S-ring, T, which has a universal isomorphism between $P_1\otimes_S T$ and $P_2\otimes_S T$, by taking the centraliser of the copy of $M_n(k)$ in the first factor of this coproduct. We shall use the symbol $S_k\langle\alpha,\alpha^{-1}:P_1 \to P_2\rangle$ for this ring.

<u>Theorem 2.16</u> Let S be a k-algebra and let P_1 and P_2 be f.g. projectives over S. Then $T = S_k\langle\alpha,\alpha^{-1}:P_1 \to P_2\rangle$ has the same global or weak dimension as S except when S has global or weak dimension 0. Here T may have dimension 0 or 1. $P_{\oplus}(T)$ is isomorphic to the quotient of $P_{\oplus}(S)$ by the relation $[P_1] = [P_2]$.

Proof: We use the notation of the preceding discussion.

$$M_n(T) \cong M_n(S_k\langle\alpha,\alpha^{-1}:P_1 \to P_2\rangle) \cong M_2(k)\times k \underset{k\times k\times k}{\sqcup} M_n(S) .$$

The global or weak dimension of $M_n(T)$ must equal that of $M_n(S)$ if this not equal to 0, by 2.10. If it equals 0, however, it can be at most 1; we see that it may be 1 by adjoining a universal isomorphism between k^1 and k^1 over k, obtaining the Laurent polynomial ring $k[t,t^{-1}]$. On the other hand, it may be 0, as we find for the ring $k \times M_2(k)$, where we adjoin a universal isomorphism between $k \times 0$ and $0 \times e_{11}M_2(k)$, obtaining $M_3(k)$.

For the last statement of the theorem, we note that $P_\oplus(M_2(k) \times k)$ is the quotient of $P_\oplus(k \times k \times k)$ by the relation $[k \times 0 \times 0] = [0 \times k \times 0]$; since $P_\oplus(M_n(T))$ is the commutative monoid coproduct of $P_\oplus(M_2(k) \times k)$ and $P_\oplus(M_n(S))$ amalgamating $P_\oplus(k \times k \times k)$, we find that the result holds for $P_\oplus(M_n(T))$, when we adjoin universal isomorphisms between $e_1M_n(S)$ and $e_2M_n(S)$. The result holds for $P_\oplus(T)$ by Morita equivalence.

Next, we deal with the adjunction of a universal idempotent map on a f.g. projective P. If we adjoin a universal idempotent map to the free module of rank 1, over the field k, we obtain the ring $k \times k$. Therefore, if the idempotent e in $M_n(S)$ satisfies $e^nS \cong P$, the S-ring that is a k-algebra with a universal idempotent map on P, $S_k\langle e:P \to P,\ e = e^2\rangle$ is obtained by taking the centraliser of the matrix units in the ring coproduct $M_n(S) \underset{k \times k}{\sqcup} k \times k \times k$, where $k \times k$ maps to $M_n(S)$ by $(a,b) \to ae + b(1-e)$, and $k \times k$ embeds in $k \times k \times k$ by $(a,b) \to (a,a,b)$.

Once again, we may deduce homological information:

Theorem 2.17 The global or weak dimension of $T = S_k\langle e:P \to P,\ e = e^2\rangle$ is equal to that of S except, possibly, when the dimension of S is 0, where T may have dimensions 0 or 1. $P_\oplus(S)$ embeds in $P_\oplus(T)$, and there are two more generators $[Q_1]$, and $[Q_2]$ subject to the relation $[Q_1] + [Q_2] = [P \otimes_S T]$.

Proof: This may be left to the reader since it in no way differs from that of the last two theorems.

It is clear that these constructions may be put together in a number of interesting ways. One construction of some interest is the adjunction to k to a pair of matrices ${}^mA^n$ and ${}^nB^m$ such that $AB = I_m$. We call this ring, R. This may be obtained by adjoining universal an idempotent n by n matrix, E, to k and then adjoining to this ring a universal isomorphism between the image of E and the free module of rank m. We are able to deduce that this is a hereditary ring and that it has exactly one new indecomposable projective, Q, up to isomorphism, and $Q \oplus R^m \cong R^n$ is the only relation that it satisfies.

Our final construction is a k-algebra that represents the functor $\mathrm{Hom}_{k\text{-alg}}(S, M_n(\ ))$; following Bergman, we call this ring $k\langle S \to M_n\rangle$.

<u>Theorem 2.18</u> Consider the ring T which is the centraliser of the first factor in the ring coproduct $M_n(k) \sqcup_k S$. Then $k\langle S \to M_n\rangle \cong T$.

Proof: Given a homomorphism $\emptyset : T \to A$, we have homomorphisms

$$S \to M_n(k) \sqcup_k S \cong M_n(T) \to M_n(A)$$

which determines a homomorphism $\emptyset_1 : S \to M_n(A)$. Conversely, if we have a map $g : S \to M_n(A)$, we have a map $g' : M_n(k) \sqcup_k S \to M_n(A)$ that sends $M_n(k)$ to $M_n(k)$ centralising A in $M_n(A)$ and acts as g on S. Restricting to the centraliser of the matrix units determines a map $g : T \to A$. These processes are mutually inverse, so they demonstrate the natural equivalence of $\mathrm{Hom}_{k\text{-alg}}(T,\ )$ and the functor $\mathrm{Hom}_{k\text{-alg}}(S, M_n(\ ))$.

Before we describe the homological properties of this construction, we consider some of its interesting ring-theoretic properties.

<u>Theorem 2.19</u> $k\langle S \to M_n\rangle$ is a domain, and the group of units is just $k^{\times}$.

Proof: If a,b are in $k\langle S \to M_n\rangle$ and $ab = 0$, we may regard a and b as endomorphisms of the induced projective module, $P \otimes_{M_n(k)} (M_n(k) \sqcup_k S)$, over the ring $M_n(k) \sqcup_k S$, where P is the simple module over $M_n(k)$. We write $R = M_n(k) \sqcup_k S$ in the following.

By 2.14, we have a commutative diagram:

$$\begin{array}{ccccc}
P\otimes_{M_n(k)}R & \xrightarrow{\ b\ } & P\otimes_{M_n(k)}R & \xrightarrow{\ a\ } & P\otimes_{M_n(k)}R \\
\downarrow & & \cong & & \uparrow \\
Q_1\otimes_{M_n(k)}R & \xrightarrow{\ \beta\otimes R\ } & P\otimes_{M_n(k)}R & \xrightarrow{\ \alpha\otimes R\ } & Q_2\otimes_{M_n(k)}R
\end{array},$$

where β, α are defined over $M_n(k)$ and compose to zero.

We know that the middle term of the bottom row is $P \otimes_{M_n(k)} R$, since this is the only representation of it as an induced module.

However, the only way that a pair of maps over $M_n(k)$, $\beta: Q_1 \to P$, and then $\alpha: P \to Q_2$ can compose to zero for one of the maps to be zero itself, which, in turn, implies that one of a and b are 0.

The result on units is a consequence of 2.3. If we have a unit in $k\langle S \to M_n \rangle$, it defines an automorphism of $P \otimes_{M_n(k)} R$; all basic transfers and transvections must be the identity map on $P \otimes_{M_n(k)} R$; so, by 2.3, the group of automorphisms of $P \otimes_{M_n(k)} R$ over R must be the group of automorphisms of P over $M_n(k)$, which is simply $k^\times$.

It is clear that the same argument shows that if $M_n(T) \cong M_n(D) \sqcup_D R$, where R is a D-ring, D is a skew field, and the isomorphism sends the obvious set of matrices to the obvious set of matrices, then T is a domain and the group of units of T is just $D^\times$.

This argument allows us to determine the global dimension of a ring coproduct over a skew field.

<u>Theorem 2.20</u> Let R_0 be a skew field and let $R = \sqcup_{R_0} R_\lambda$, where each R_λ is an R_0-ring; then, the global of R equals the supremum of the global dimensions of R except when they are all zero; then it becomes 1.

Proof: By 2.10, we have this result except possibly when the rings are all semi-simple. Consider the ring of endomorphisms of an induced projective module $P \otimes_{R_\lambda} R$, where P is a simple module for the semisimple ring R_λ. The group of units is simply the group of automorphisms of the module and by the same argument as in the last theorem we see that this is just the group of automorphisms of P over R_λ. However, the ring of endomorphisms of $P \otimes_{R_\lambda} R$ is larger than the ring of endomorphisms of P over R_λ, so that this ring of endomorphisms is not a skew field. Therefore, R cannot be a semisimple artinian ring since over a semisimple artinian ring, an indecomposable projective module is simple, and so its ring of endomorphisms must be a skew field.

We see that the global dimension of $k\langle S \to M_n \rangle$ is equal to that of S by 2.10, except when S is semisimple artinian in which case we may apply 2.20 to show that it is equal to 1.

It is interesting to examine some of these constructions in the category of commutative algebras. We shall not provide detailed proofs since this last part is meant to be only illustrative.

We begin with a commutative ring C and a couple of f.g. projectives P_1 and P_2 over C. We wish to find a commutative C-algebra $C[\alpha:P_1 \to P_2]$ that represents the functor on the category of C-algebras $A \to \mathrm{Hom}_{A\text{-mod}}(P_1 \otimes_C A, P_2 \otimes_C A)$. The functor $\mathrm{Hom}_A(P_1 \otimes_C A, P_2 \otimes_C A)$ is naturally equivalent to the functor $\mathrm{Hom}_A((P_1 \otimes_C P_2^{\times}) \otimes_C A, A)$, where $P_2^{\times}$ is the dual of P_2; but it is clear that the C-algebra representing this functor is just the symmetric algebra on the module $P_1 \otimes_C P_2$ over C. It is an immediate consequence that this algebra is <u>geometrically regular over C</u>; that is, for each prime ideal p of C, $C[\alpha:P_1 \to P_2] \otimes_C \overline{Q(C/p)}$ is a regular ring where $\overline{Q(C/p)}$ is the algebraic closure of the ring of quotients of C/p.

We shall find that this holds for each construction examined and we shall be able to obtain each construction in a sufficiently explicit form to be able to determine the dimension of each geometric fibre over C. For example, in the above example, the fibre over the prime p is $\overline{Q(C/p)}[x_{ij}: i = 1 \to m_1,\ j = 1 \to m_2]$, where m_i is the local rank of P_i at p.

We examine next the construction of adjoining a universal isomorphism from P_1 to P_2. This is only possible when the local rank of P_1 is the same as that of P_2. If such is the case, P_1 becomes isomorphic to P_2 on a Zariski cover and by further refinement we see that there is a Zariski cover which is the union of finitely many open and closed subspaces such that the restriction of P_1 to each of these subsets is isomorphic to the restriction of P_2 and both are isomorphic to free modules on each subspace. If on a particular subspace the restrictions have local rank n, then the space of all isomorphisms is a principal homogeneous space for GL_n. Therefore, we see that the fibre at a prime of C of the spectrum of the ring representing the space of isomorphisms of P_1 with P_2 is just $GL_n(\overline{Q(C/p)})$, where n is the local rank of P_i.

If we wish to adjoin an idempotent map $e:P \to P$ to C, it is reasonable to specify the local rank of the idempotent also, which must be constant on connected components and dominated by that of P. We may as well restrict to a situation where the local rank of P is constant and so is that of the idempotent at n and m respectively. So we wish to represent idempotents of rank m in $\mathrm{End}_A(P\otimes_C A)$. Over C, there is a Zariski cover on which P becomes free of rank n. In a particular algebra, any two idempotents of rank m in $M_n(A)$ present free modules with free complement on a suitable Zariski cover, so that with respect to the Zariski topology all these elements are conjugate. Moreover, the centraliser of an idempotent whose image is free of rank m and whose complement is free of rank (n-m) is just $M_m(A)\times M_{n-m}(A)$; the consequence of this is that the set of idempotents of rank m in $\mathrm{End}_A(P\otimes_C A)$ becomes on a Zariski cover the functor $GL_n/GL_m\times GL_{n-m}$. Since this last functor is geometrically regular, so is the one we are interested in, and it is easy to check that the dimension of a fibre is 2mn.

3 PROJECTIVE RANK FUNCTIONS ON RING COPRODUCTS

The purpose of this chapter is to investigate how projective rank functions behave under the coproduct construction. On the way we shall give an application of this theory to the problem of accessibility for f.g. groups. Also, we shall prove some interesting results on the behaviour of the number of generators of an induced module over a coproduct. The main applications of the results of this chapter will be to hereditary rings in the following chapters.

The Generating Number on Ring Coproducts

When investigating partial projective rank functions on a ring coproduct, $R = R_1 \sqcup_{R_0} R_2$, where R_0 is a semisimple artinian ring and R_i is a faithful R_0-ring, it is sensible to assume that it is defined on the image of $K_0(R_0)$ in $K_0(R)$. In this situation, we have the following lemma:

Lemma 3.1 The partial projective rank functions on a ring coproduct $R = R_1 \sqcup_{R_0} R_2$ where R_0 is a semisimple artinian ring and R_i is a faithful R_0-ring, that are defined on the image of $K_0(R_0)$ in $K_0(R)$ are given by pairs of partial projective rank functions (ρ_1, ρ_2) where ρ_i is a partial projective rank function on R_i defined on the image of $K_0(R_0)$ such that they induce the same projective rank function on R_0.

Proof: This is clear since by 2.12, we have a push-out diagram:

$$\begin{array}{ccc} K_0(R_0) & \to & K_0(R_1) \\ \downarrow & & \downarrow \\ K_0(R_2) & \to & K_0(R) \end{array}$$

We shall usually describe a partial projective rank function of this type by the pair (ρ_1,ρ_2) of partial projective rank functions where ρ_i is defined on R_i.

We wish to show that the inner projective rank of a map $\alpha:P \to Q$ defined over R_1 with respect to a rank function ρ is just the same as the inner rank of $\alpha \otimes_{R_1} R: P\otimes_{R_1} R \to Q\otimes R_1 R$ over $R = R_1 \sqcup_{R_0} R_2$ with respect to the projective rank function (ρ_1,ρ_2). In order to do this, we need to investigate how the generating number with respect to a rank function behaves under the coproduct. Before we do this we present a result that is independent of rank functions.

In chapter 1, we have already encountered the notion of a ring with unbounded generating number; that is, there can be no equation of the form ${}^mR_1 \cong {}^{m+1}R_1 \oplus P$. If however such an equation holds, a similar equation holds for all R_1-rings, and in particular for a ring coproduct, $R = R_1 \sqcup_{R_0} R_2$. So the number of generators of a module induced from R_2 may well change. We show that this is the only way that things can go wrong.

Theorem 3.2 Suppose that R_0 is a skew field, and R_2 has unbounded generating number, where R_1 and R_2 are R_0-rings. Then the minimal number of generators over R_1 of the f.g. module M is equal to the minimal number of generators of the module $M\otimes_{R_1} R$ over R.

Proof: Suppose that $M\otimes_{R_1} R$ may be generated by m elements over R; then there is a surjection of induced modules $\alpha: {}^mR_1\otimes_{R_1} R \to M\otimes_{R_1} R$. By 2.3, there is an isomorphism $\beta: P_1\otimes_{R_1} R \oplus P_2\otimes_{R_1} R \to {}^mR_1\otimes_{R_1} R$, which is a finite composition of free transfers and transvections such that the composite $\alpha\beta$ is an induced map; that is, it maps P_1 onto M and P_2 to 0. We shall show that P_1 is an m-generator module and so is M.

We recall from theorem 1.4 that if R_2 has unbounded generating number, we have a well-defined partial rank function ρ_2 on stably free R_2-modules. Since any stably free module has non-negative rank, the rank of a free summand of a stably free module P is bounded by the rank $\rho_2(P)$.

If after a series of free transfers and transvections we have passed from ${}^mR_1\otimes_{R_1} R$ to $P_1\otimes_{R_1} R \oplus P_2\otimes_{R_1} R$, we shall see that $P_1 \oplus {}^{\rho_2(P_2)}R_1 \cong {}^mR_1$. We assume that this is true at the nth stage; if our next step is a transvection, it alters nothing; if it is a free transfer from P_1 to P_2 it is still true; if it is a free transfer from P_2 to P_1 we can transfer at most $\rho_2(P_2)$ factors, so it is still true. It is true at the beginning so by induction it is true at the end.

We have shown the hard direction of our theorem; the other is trivial.

We pass to the more technical versions we need for partial projective rank functions in general.

<u>Theorem 3.3</u> Let $R = R_1 \underset{R_0}{\sqcup} R_2$, where R_0 is a semisimple artinian ring and each R_i is a faithful R_0-ring; let ρ_1, ρ_2 be partial projective rank functions for R_1 and R_2 respectively, defined and agreeing on $K_0(R_0)$ in $K_0(R_i)$. Let M_1 and M_2 be f.g. R_1 and R_2 modules respectively with generating numbers m_1 and m_2 respectively with respect to ρ_1 and ρ_2. Then the generating number of $M_1 \otimes_{R_1} R \oplus M_2 \otimes_{R_1} R$ with respect to (ρ_1, ρ_2) is $m_1 + m_2$.

Proof: Certainly, it is at most $m_1 + m_2$.

Conversely, suppose that we have a surjection over R, $(P_1 \otimes_{R_1} R) \oplus (P_2 \otimes_{R_2} R) \to (M_1 \otimes_{R_1} R) \oplus (M_2 \otimes_{R_2} R)$, where $\rho_2(P_i)$ is defined. By 2.3, after a series of basic transfers and transvections on $P_1 \otimes_{R_1} R \oplus P_2 \otimes_{R_2} R$, we obtain an induced surjection:

$$P_1' \otimes_{R_1} R \oplus P_2' \otimes_{R_2} R \to M_1 \otimes_{R_1} R \oplus M_2 \otimes_{R_2} R$$

Since ρ_i is defined on the image of $K_0(R_0)$ in $K_0(R_i)$ and agree there, the basic transfers produce modules on which ρ_i is defined. Since P_i' maps onto M_i, $\rho_i(P_i') \geq m_i$, so the rank of $P_1 \otimes_{R_1} R \oplus P_2 \otimes_{R_2} R$ with respect to (ρ_1, ρ_2) must be at least $m_1 + m_2$.

At this point, it is of some interest to show how this theory may be applied to provide a fairly simple proof of a result of (Linnell 83) on accessibility for f.g. groups.

We begin with a summary of the background to the problem, and refer the reader to (Dicks 80) for the details of the subject.

Let X be a finite connected graph with edge set E and vertex set V; we allow the beginning τe and end ιe of an edge to be the same vertex. We label the edges and vertices of X with groups, and for each edge e we have embeddings $G \to G_{\tau e}$ and $G_e \to G_{\iota e}$; if $\tau e = \iota e$, these embeddings may well be different. This is known as a graph of groups; we associate to this a group, the fundamental group of a graph of groups in the following way.

Let T be a maximal subtree of X, which we regard as a subtree of groups by the appropriate labelling of the elements of T; let v be a vertex of T with only one edge, e, incident with it, so $v = \tau e$, without loss of generality. We pass to the tree with one fewer vertex, T', obtained by omitting e and τe, which we label by the same groups as before except for the vertex, e, which we label with the group coproduct of $G_{\tau e}$ and $G_{\iota e}$ amalgamating G_e, $G_{\tau e} \underset{Ge}{*} G_{\iota e}$. By induction, we eventually reach a single point with a group G_T associated to it, with a specified homomorphism $G_v \to G_T$ for each vertex v such that $G_{\tau e} \cap G_{\iota e} = G_e$ for each edge e in T; clearly, G_T is universal with respect to this property.

For each edge e in $X - T$, we have two embeddings of G_e in G_T, $G_e \to G_{\tau e} \subseteq G_T$ and $G_e \to G_{e} \subseteq G_T$; so we form the multiple HNN extension of G_T over all edges of $X - T$ with respect to these pairs of embeddings. The resulting group, G_X, is the fundamental group of our graph of groups; it can be shown that it is independent of the maximal subtree chosen. Also, if X' is a full connected subgraph of X, we may express G_X as the fundamental group of a graph of groups on X, where X is the graph obtained from X by shrinking X' to a point. The groups associated to the elements of X are the same as those in X for those that are not affected and that associated to the point to which we shrunk X' is the group $G_{X'}$.

In the following, we shall assume that edges groups are finite, and also that if there is an edge e such that $G_e = G_{\tau e}$ or $G_e = G_{\iota e}$, then $\tau e = \iota e$.

Given a f.g. group, G, we are interested in the various ways of expressing G as the fundamental group of a graph of groups with finite edge groups. It is possible to show that if G is the fundamental group of a graph of groups on X_1 and X_2, then these two representations have a common refinement, that is, there is a representation of G as the fundamental group of a graph of groups on a graph X such that the two preceding representations arise by collapsing suitable subgraphs of X to points. So the question arises whether there is a representation from which all others arise by collapsing suitable subgraphs. If there is, the group is said to be accessible. This is equivalent to finding a representation where all the vertex groups are neither HNN extensions over some finite group nor non-trivial coproducts over some finite group. Since it is also easy to see that the number of generators of the fundamental group of a graph of groups on a graph X is at least $E(X) - V(X) + 1$ (consider the homomorphism to the fundamental graph of groups on the graph X where all vertex groups and edge groups are trivial, which is the free group on the stated number of generators), we know that if the vertex groups keep on decomposing they must eventually decompose only as coproducts over finite groups, so that we shall only consider this possibility. It is clear that this cannot happen for f.g. torsion-free groups since the number of generators of $H_1 * H_2$ is the sum of the number of generators of H_1 and H_2 by the Grushko-Neumann theorem. In order to prove a more general theorem we need to find a substitute for the number of generators that grows in a satisfactory way for the coproduct with amalgamation over a finite group. Linnell (83) was able to do this for those groups whose subgroups are of bounded order; we shall present his theorem, using partial rank functions instead of the analysis he used.

We noted in the first chapter, that on every group ring in characteristic 0 there is a faithful projective rank function on f.g. projectives induced by the trace function. We are interested in the subgroup $K_0^f(KG)$ of $K_0(KG)$ which is generated by the f.g. projectives induced up from the subgroups of finite order, where K is a field of characteristic 0. We shall denote by ρ^f the partial projective rank function induced by the trace on $K_0^f(KG)$. If $G = G_1 \underset{G_0}{*} G_2$, then $KG \cong KG_1 \underset{KG_0}{\sqcup} KG_2$; if F is a finite group, then KF is semisimple artinian and we should like to show that ρ^f is just the partial projective rank function (ρ_1^f, ρ_2^f).

<u>Lemma 3.4</u> Let $G = G_1 \underset{F}{*} G_2$ where F is a finite group; then $K_0^f(KG)$ is just the subgroup of $K_0^f(KG)$ generated by the images of $K_0^f(KG_i)$. Consequently, ρ^f is the partial projective rank function (ρ_1^f, ρ_2^f).

Proof: If P is a f.g. projective induced up from some finite subgroup H of G, we know that H is a conjugate of some subgroup H' of either G_1 or G_2; consequently, P is isomorphic to a projective induced from H' and must lie in the image of $K_0^f(KG_1)$ or $K_0^f(KG_2)$. So $K_0^f(KG)$ is generated by the images of $K_O^f(KG_1)$ and $K_O^f(KG_2)$.

Since ρ_1^f and ρ_2^f agree on the image of $K_O(KF)$, lemma 3.1 shows that our last statement is true.

This result allows us to show Linnell's theorem; we shall use the generating number with respect to ρ of the augmentation ideal of G in KG to measure the size of G.

Theorem 3.5 Let G be a f.g. group such that the number of elements in finite subgroups of G is bounded; then G is accessible.

Proof: We recall that it is sufficient to show that we cannot keep on decomposing such groups as a group coproduct amalgamating a finite subgroup in a non-trivial way.

If $G \cong G_1 \underset{F}{*} G_2$, where $|F| < \infty$, we know that

$$\omega G \cong (\omega G_1)G \oplus (\omega G_2/\omega FG_2)G ,$$

where ωH is the augmentation ideal of a group H in KH.

$KG \cong KG_1 \underset{KF}{\sqcup} KG_2$ and it is clear from the above equation that ωG is an induced module. By 3.3 and 3.4, we see that

$$g.\rho^f(\omega G) = g.\rho_1^f(\omega G_1) + g.\rho_2^f(\omega G_2/\omega FG_2),$$

and it is clear that if m is the bound on the order on finite subgroups of G, then $g.\rho_2^f(\omega G_2/\omega FG_2) \geq \frac{1}{m}$.

Consequently, if $g.\rho^f(\omega G) = q$, there is no decomposition of G as the vertex group of a tree of groups with finite edge groups having more than $(m.q)$ vertices, which proves Linnell's theorem.

Sylvester projective rank functions on ring coproducts

We return to the general theory. Consider a ring homomorphism $\emptyset : R \to S$; suppose that we have a projective rank function on S, ρ_S, inducing a projective rank function ρ_R on R. We say that the map $\emptyset$ is honest with respect to the projective rank functions ρ_R and ρ_S if for any map $\alpha : P \to Q$ in the category of f.g. projectives over R, $\rho_R(\alpha) = \rho_S(\alpha \otimes_R S)$.

This clearly reduces to the usual notion of honesty for firs (Cohn 71, p.264).

<u>Theorem 3.6</u> Let $R = R_1 \sqcup_{R_0} R_2$, where R_0 is a semisimple artinian ring and each R_i is a faithful R_0-ring. Let ρ_i be projective rank functions on R_i that induce the same projective rank function on R_0. Let $\rho = (\rho_1,\rho_2)$ be the projective rank function they define on R. Then the embedding $R_1 \to R$ is honest with respect to ρ_1 and ρ.

Proof: Let $\alpha : P \to Q$ be a map between f.g. projectives over R_1. It is clear that $\rho_1(\alpha) \geq \rho(\alpha \otimes_{R_1} R)$.

So let M be a f.g. R-module such that $\alpha(P)\otimes_{R_1} R \subseteq M \subseteq Q\otimes_{R_1} R$. By 1.10, we need to show that $g.\rho(M) \geq \rho_1(\alpha)$.

By 2.8, the decomposition of M given by 2.7 has the form

$$M = (M_0 \otimes_{R_0} R) \oplus (M_1 \otimes_{R_1} R) \oplus (M_2 \otimes_{R_2} R),$$

where $\alpha(P) \subseteq M_1$. By 3.3, the generating number of M with respect to ρ is at least the generating number of M_1 with respect to ρ_1.

By 2.1, the embedding of R_1-modules $Q \subseteq Q\otimes_{R_1} R$ splits; so consider the image of M_1, M', in Q under the splitting map. It contains the image of P under α, so its generating number with respect to ρ_1 is at least $\rho_1(\alpha)$. We have shown $g.\rho(M) \geq g.\rho_1(M_1) \geq g.\rho_1(M') \geq \rho_1(\alpha)$, so $\rho(\alpha\otimes_{R_1} R) = \rho_1(\alpha)$, as we wished to show.

This result will be of great use to us later; for the present, we give a couple of interesting consequences.

<u>Theorem 3.7</u> Let $R = R_1 \sqcup_{R_0} R_2$, where R_0 is semisimple artinian, and each R_i is a faithful R_0-ring. Let $\rho = (\rho_1,\rho_2)$ be a projective rank function on R. Then ρ is a Sylvester projective rank function if and only if each ρ_i is a Sylvester projective rank function.

Proof: Suppose that ρ_i is a Sylvester projective rank function on R_i. Consider a couple of maps $\alpha : P \to P'$, and $\beta : P' \to P''$ defined over R such that $\alpha\beta = 0$. Then by 2.13, we have a commutative diagram:

$$\begin{array}{ccccc} P & \xrightarrow{\alpha} & P' & \xrightarrow{\beta} & P'' \\ \downarrow & & \cong & & \uparrow \\ \underset{i}{\oplus} P_i \otimes_{R_i} R & \xrightarrow{\oplus \alpha_i \otimes R} & \underset{i}{\oplus} P'_i \otimes_{R_i} R & \xrightarrow{\oplus \beta_i \otimes R} & \underset{i}{\oplus} P''_i \otimes_{R_i} R \end{array} ,$$

where α_i, β_i are maps defined over R_i such that $\alpha_i \beta_i = 0$.

By the law of nullity, $\rho_i(\alpha_i) + \rho_i(\beta_i) \leq \rho_i(P')$; summing, we find that $\rho(\alpha) + \rho(\beta) \leq \sum_i \rho_i(\alpha_i) + \rho_i(\beta_i) \leq \rho(P')$, so that ρ is a Sylvester projective rank function too.

Conversely, suppose that ρ is a Sylvester projective rank function, and let $\alpha : P \to P'$, and $\beta : P' \to P''$ be maps between f.g. projectives over R_1 such that $\alpha\beta = 0$. Since ρ is a Sylvester projective rank function, $\rho(\alpha \otimes_{R_1} R) + \rho(\beta \otimes_{R_1} R) \leq \rho(P' \otimes_{R_1} R)$; but, by the last theorem, $\rho(\alpha \otimes_{R_1} R) = \rho_1(\alpha)$ for any maps defined over R_1. So the projective rank function ρ_1 is Sylvester.

<u>Theorem 3.8</u> If $R = R_1 \underset{R_0}{\sqcup} R_2$, where R_0 is a skew field, then R is a Sylvester domain if and only if R_1 and R_2 are Sylvester domains.

Proof: This is an immediate consequence of the last theorem when we notice that all the f.g. projectives are free of unique rank over R if and only if this is true for each R_i.

We have seen in theorem 3.6 that if $R = R_1 \underset{R_0}{\sqcup} R_2$ for semisimple artinian R_0, and faithful R_0-rings R_1 and R_2 and $\rho = (\rho_1, \rho_2)$ is a projective rank function on R, that full maps with respect to ρ_1 remain full maps with respect to ρ when they are induced up to R. In the case where the rank functions are all Sylvester, we are able to prove that the factorisations of a full map induced up from R_1 to R over R essentially come from R_1.

<u>Theorem 3.9</u> Let $R = R_1 \underset{R_1}{\sqcup} R_2$, where R_0 is semisimple artinian and each R_i is a faithful R_0-ring. Assume that $\rho = (\rho_1, \rho_2)$ is a Sylvester faithful projective rank function on R (and, in consequence, the same holds for ρ_i on R_i). Then, if $\alpha \otimes_{R_1} R = \beta\gamma$ is a factorisation of a full map as a product

of full maps over R, there is an invertible map, ε, such that $\beta\varepsilon$ and $\varepsilon^{-1}\gamma$ are defined over R_1.

Proof: Suppose that we have the factorisation over:

$$\begin{array}{ccc} P\otimes_{R_1} R & \xrightarrow{\alpha\otimes_{R_1} R} & Q\otimes_{R_1} R \\ & {\scriptstyle\beta}\searrow \quad \nearrow{\scriptstyle\alpha} & \\ & P' & \end{array}$$

where β,γ are full maps with respect to ρ, the rank function on R. So $\rho(P') = \rho_1(P) = \rho_1(Q)$. γ must be an embedding, since it is a full map with respect to a faithful Sylvester rank function. Since $\gamma(P') \supseteq \alpha(P)$, we know by 2.8 that the decomposition of $\gamma(P')$ given by 2.7 takes the form

$$\gamma(P') = M_0\otimes_{R_0} R) \oplus (M_1\otimes_{R_1} R) \oplus (M_2\otimes_{R_2} R),$$

where $\alpha(P) \subseteq M_1$.

If one of M_0 or M_2 is not zero, its rank with respect to the relevant rank function is not zero; consequently, $\rho_1(M_1) < \rho_1(P)$ and so $\alpha\otimes_{R_1} R$ could not be full since its image lies inside $M_1\otimes_{R_1} R$. So $\gamma(P') = M_1\otimes_{R_1} R$.

Now all we need to do is to show that $M_1 \subseteq Q$, for then we take ε to be the identification of P' with $M_1\otimes_{R_1} R$, and we see that $\beta\varepsilon$ is the induced map $\bar{\beta}\otimes_{R_1} R$, where $\bar{\beta}$ is the map $P \to \alpha(P) \subseteq M_1$ and $\varepsilon^{-1}\gamma$ is the induced map $\bar{\gamma}\otimes_{R_1} R$, where $\bar{\gamma}$ is the inclusion of M_1 in Q.

Since $\rho_1(P) = \rho_1(M_1)$ and $\bar{\beta}$ is a factor of the full map α, we see that $\bar{\beta}$ is a full map. We consider the map over R_1 $\delta: M_1 \subseteq Q\otimes_{R_1} R \to (Q\otimes_{R_1} R)/Q$, and notice that $\bar{\beta}\delta = 0$; hence, since the rank function is faithful $\delta = 0$, so that $M_1 \subseteq Q$ as desired.

4. UNIVERSAL LOCALISATION

One of the first constructions developed in the theory of rings was localisations for commutative rings as a way of passing from a commutative domain to its field of fractions. When the theory of non-commutative rings was developed, it was noticed that an analogue of this was possible for suitable sets of elements in a ring provided that they satisfied the Ore condition. This method was shown to be of particular importance when Goldie showed that for a prime Noetherian ring the set of non-zero-divisors of the ring satisfied the Ore condition and the Ore localisation at this set was a simple artinian ring.

One construction that was considered but rejected on the grounds that at the time nothing could be proved about it was the construction of adjoining universally the inverses to a subset of elements of the ring. Of course, the Ore localisation is a special case of this construction. However, in studying the homomorphisms from rings to skew fields, Cohn was forced to study a generalisation of this construction. He showed that the set of a matrices over a ring R that become invertible under an epimorphism to a skew field F determine the epimorphism; more specifically, the ring obtained by adjoining universally the inverses to these matrices is a local ring L whose residue skew field is F. Of course, the homomorphism $R \to L \to F$ is our original epimorphism. For the first time, therefore, the construction of adjoining universally the inverses of some set of matrices required serious consideration; this was taken a step further by Bergman who considered the construction of adjoining universally the inverses to a set of maps between f.g. projectives.

By now, there is a powerful but secret body of knowledge about this construction that is of fundamental importance for the study of homomorphisms to simple artinian rings. The purpose of this chapter is to present this theory which has been substantially simplified and extended recently.

First, we shall give a simple existence proof for the construction;

this allows us to develop standard expressions for the elements of the universal localisation together with a simple description of the equivalence relation on these expressions given by their equality in the universal localisation. This is based on and generalises work of Cohn (71) and Malcolmson (83). Then we generalise an argument of Dlab and Ringel (84) to find simple homological information on the universal localisation; this allows us to give a very simple proof of a result of Dicks and Bergman (78); a universal localisation of a right hereditary ring must be right hereditary. Also, there is a short discussion of an interesting way to study the ring coproduct in terms of the universal localisation of a suitable ring. In the last section, we prove an exact sequence in algebraic K-theory that generalises the exact sequence of Bass and Murthy (67) for Ore localisation.

Normal forms for universal localisation

Let R be a ring, and let $\underline{P}(R)$ be the category of f.g. left projective modules over R; let Σ be some set of maps in the category; we wish to construct a ring R_Σ that is universal with respect to the property that the maps $R_\Sigma \otimes_R \alpha$ for $\alpha \in \Sigma$ are invertible.

Theorem 4.1 Let R be a ring and Σ be a set of maps between f.g. left projectives. Then there is an R-ring R_Σ universal with respect to the property that every element $R_\Sigma \otimes_R \alpha$ for $\alpha \in \Sigma$ has an inverse.

Proof: We use the observation that if we take the category of f.g. projectives over a ring R, and adjoin a set of maps between various objects subject to some set of relations and consider the additive category generated by these maps and relations and our original category, then this is a full subcategory of the category of f.g. projectives over the endomorphism ring of the object that was the free module of rank 1 over R in our initial category. In our case, we adjoin a set of maps $\bar{\alpha}:Q \to P$ for each map $\alpha:P \to Q$ in Σ and adjoin the relations that $\alpha\bar{\alpha} = I_P$, and $\bar{\alpha}\alpha = I_Q$ for all α in Σ.

Clearly the map from R to the endomorphism ring of the object that was the free module of rank 1 over R makes this ring into an R-ring with the correct universal property. We shall call this ring R_Σ. It is clear that the category of modules that we have constructed consists of the f.g. projectives over R_Σ that are induced from such projectives over R.

Now that we have shown the existence of this construction, we should note that there is no difference between adjoining universally the inverses of maps between left f.g. projectives or between right f.g. projectives since we adjoin an inverse to $\alpha:P \to Q$ if and only if we adjoin an inverse to $\alpha^x:Q^x \to P^x$.

We have shown the existence of the universal construction, but this does not allow us to show in special cases that the ring we have is non-trivial; therefore, we need a concrete representation of the elements of the ring together with a way for determining when to such representations are equal; this is our next goal.

First of all, we show that every map in the category of induced f.g. projective modules over a universal localisation R_Σ of R may be represented in the form $f\gamma^{-1}g$, where f, g and γ are maps defined over R and $\gamma = \begin{pmatrix} \alpha_1 & & * \\ & \ddots & \\ 0 & & \alpha_n \end{pmatrix}$ where each $\alpha_i \in \Sigma$. This is clearly true for induced maps and for the maps $\bar{\alpha}:Q \to P$ for each $\alpha:P \to Q$, where α is in Σ.

If $f_1\gamma_1^{-1}g_1$ and $f_2\gamma_2^{-1}g_2$ are both maps from $R_\Sigma \otimes_R P$ to $R_\Sigma \otimes_R O$, then we find the equation

$$f_1\gamma_1^{-1}g_1 - f_2\gamma_2^{-1}g_2 = (f_1 \quad f_2)\begin{pmatrix} \gamma_1 & 0 \\ 0 & \gamma_2 \end{pmatrix}^{-1}\begin{pmatrix} g_1 \\ -g_2 \end{pmatrix} .$$

If $f_1\gamma_1^{-1}g_1:R_\Sigma \otimes_R P_1 \to R_\Sigma \otimes_R P_2$ and $f_2\gamma_2^{-1}g_2:R_\Sigma \otimes_R P_2 \to R_\Sigma \otimes_R P_3$ are a couple of maps, then

$$f_1\gamma_1^{-1}g_1 . f_2\gamma_2^{-1}g_2 : R_\Sigma \otimes_R P_1 \to R_\Sigma \otimes_R P_3$$

is given by

$$f_1\gamma_1^{-1}g_1 . f_2\gamma_2^{-1}g_2 = (f_1 \quad 0)\begin{pmatrix} \gamma_1 & -g_1 f_2 \\ 0 & \gamma_2 \end{pmatrix}^{-1}\begin{pmatrix} 0 \\ g_2 \end{pmatrix}$$

We see that every map must have this form since all induced maps are got by successive composition or taking the differences of maps previously found. The reason for considering such representations of the maps is that we are able to write down a criterion for two such expressions to be equal maps in R_Σ.

<u>Theorem 4.2 (Malcolmson's criterion)</u> Let R be a ring and Σ a collection of maps between f.g. projectives over R. Then every map between induced projectives over R_Σ has the form $f\gamma^{-1}g$ for maps f, g, and γ defined over R, and $\gamma = \begin{pmatrix} \alpha_1 & & * \\ & \ddots & \\ 0 & & \alpha_n \end{pmatrix}$ for suitable $\alpha_i \in \Sigma$; further,

$f_1\gamma_1^{-1}g_1 = f_2\gamma_2^{-1}g_2$ if and only if there is an equation where all maps are defined over R.

$$\left(\begin{array}{cccc|c} \gamma_1 & 0 & 0 & 0 & g_1 \\ 0 & \gamma_2 & 0 & 0 & g_2 \\ 0 & 0 & \gamma_3 & 0 & 0 \\ 0 & 0 & 0 & \gamma_4 & g_4 \\ \hline f_1 & f_2 & f_3 & 0 & 0 \end{array}\right) = \begin{pmatrix} \mu \\ \delta \end{pmatrix} (\nu \mid \varepsilon)$$

where $\gamma_3, \gamma_4, \mu, \nu$ all have the form $\begin{pmatrix} \alpha_1 & & * \\ & \ddots & \\ 0 & & \alpha_n \end{pmatrix}$ for suitable $\alpha_i \in \Sigma$.

We refer the reader to Malcolmson (83) for a proof in the case where all elements of Σ are matrices over R; he only considers elements of the ring R_Σ instead of all maps between induced projective modules; the transformation to this present form is purely mechanical. The only apparent difficulty arises from the fact that addition and multiplication are not defined everywhere; however, a moment's thought shows that this causes no problem.

There is another representation of the maps in the category of induced f.g. projectives over R that is often useful for performing certain calculations; this is a generalisation of techniques of Cohn.

<u>Theorem 4.3 (Cramer's rule)</u> Let R be a ring and Σ a collection of maps between f.g. left projectives. Then in the category of induced f.g. projectives over R_Σ, every map $\alpha: R_\Sigma \otimes_R P \to R_\Sigma \otimes_R Q$ satisfies an equation of the form:

$$\beta\begin{pmatrix} I & \alpha' \\ \hline O & \alpha \end{pmatrix} = \beta'$$

where β, β' are both maps defined over R, and $\beta = \begin{pmatrix} \beta_1 & & O \\ & \ddots & \\ & & \beta_n \end{pmatrix}$, where $\beta_i \in \Sigma$, I is an identity map on a suitable projective, and α' is some map over R_Σ.

Proof: We use the generator and relation construction of the category of induced f.g. projectives over R_Σ given in the preceding discussion.

Certainly, for β in Σ, the element $\overline{\beta}$ satisfies

$$\beta\,\overline{\beta} = I$$

which is of the required form; for α defined over R, the equation

$$I\alpha = \alpha$$

is of the required form.

Next, given $\alpha_i : P \to Q$ $i = 1,2$ satisfying equations:

$$\left(\beta_i \mid \beta_i'\right)\begin{pmatrix} I & \alpha_i' \\ \hline O & \alpha_i \end{pmatrix} = \left(\beta_i \mid \beta_i''\right)$$

we find that for $\alpha_1 - \alpha_2$, we have the equation

$$\begin{pmatrix} \beta_1 & \beta_1' & O & O \\ O & \beta_2' & \beta_2 & \beta_2' \end{pmatrix}\begin{pmatrix} & -\alpha_1' \\ I & -\alpha_1 \\ & -\alpha_2' \\ \hline O & \alpha_1 - \alpha_2 \end{pmatrix} = \begin{pmatrix} \beta_1 & \beta_1' & O & -\beta_1'' \\ O & \beta_2' & \beta_2 & -\beta_2'' \end{pmatrix}$$

Finally, suppose that we have equations for i = 1,2,

$$(\beta_i \mid \beta_i')\begin{pmatrix} I & \alpha_i' \\ \hline O & \alpha_i \end{pmatrix} = \left(\beta_i \mid \beta_i''\right)$$

for maps $\alpha_1 : P_1 \to P_2$ and $\alpha_2 : P_2 \to P_3$; then we construct the equation

$$\begin{pmatrix} \beta_2 & \beta_2' & 0 & \vline & 0 \\ 0 & -\beta_1'' & \beta_1 & \vline & \beta_1' \end{pmatrix} \left(\begin{array}{c|c} I & \begin{matrix} \alpha_2' \\ \alpha_2 \\ \alpha_1'\alpha_2 \end{matrix} \\ \hline 0 & \alpha_1\alpha_2 \end{array}\right) = \begin{pmatrix} \beta_2 & \beta_2' & 0 & \vline & \beta_2'' \\ 0 & -\beta_1'' & \beta_1 & \vline & 0 \end{pmatrix}$$

Since every map in the category of induced f.g. projectives may be obtained from the generators by these two processes of differences and compositions, this proves the theorem.

Of course, in the last theorem there are no good reasons why the representation of α in the equation $\beta\begin{pmatrix} I & \alpha' \\ 0 & \alpha \end{pmatrix} = \beta'$ should be in any way unique. We shall see however that Malcolmson's criterion gives us a great deal of control over such representations.

At this point, it is useful to introduce a pair of formal definitions that we need later. We say that a set of maps Σ between f.g. projectives is <u>lower multiplicatively closed</u> if $\alpha, \beta \in \Sigma$ implies that $\begin{pmatrix} \alpha & 0 \\ \gamma & \beta \end{pmatrix} \in \Sigma$ for arbitrary suitably sized γ. Similarly, we may define <u>upper multiplicatively closed</u>. A set of maps Σ is said to be <u>saturated</u> if every map between f.g. projectives over R that become invertible over R_Σ is associated over R to an element of Σ. We may define the <u>lower</u> and <u>upper multiplicative closure</u> of a set of maps in the obvious way; it is convenient to be more careful when defining a saturation of a set of maps Σ: a <u>saturation</u> of Σ is a lower and upper multiplicatively closed set of maps such that every map between f.g. projectives over R that becomes invertible over R_Σ is associated to an element of the saturation. For Cramer's rule, it is useful to use lower multiplicatively closed sets of maps and for Malcolmson's criterion it is better to use upper multiplicatively closed ones.

We note the following consequences of Cramer's rule.

<u>Corollary 4.4</u> All maps between induced f.g. projectives over R_Σ are stably associated to induced maps.

Proof: Simply look at the form of Cramer's rule.

<u>Corollary 4.5</u> All finitely presented modules over R_Σ are induced from

finitely presented modules over R.

Proof: A f.p. module is the cokernel of an m by n matrix; by 4.4, this is stably associated to an induced map between f.g. projectives, and cokernels are isomorphic for stably associated maps. So all f.p. modules over R_Σ are cokernels of induced maps. Let $R_\Sigma \otimes_R \alpha : R_\Sigma \otimes_R P \to R_\Sigma \otimes_R Q$ be an induced map; then by the right exactness of the tensor product, the cokernel of $R_\Sigma \otimes_R \alpha$ is isomorphic to $R_\Sigma \otimes_R (\mathrm{coker}\,\alpha)$ proving the result.

In particular, this applies to the f.g. projectives over R_Σ; however, we cannot conclude that all f.g. projectives over R_Σ are induced from R; for example, consider $\mathbb{Z} + 2M_2(\mathbb{Z}) \subseteq M_2(\mathbb{Z})$; all f.g. projectives are free; however, the central localisation of this ring at 2 is isomorphic to $M_2(\mathbb{Z}_2)$, and there are some new projectives in this case.

One use of 4.4 is to show that iterated universal localisations are in general universal localisations.

Theorem 4.6 Let Σ be a collection of maps between f.g. projectives over R, and let Σ' be a collection of maps between stably induced f.g. projectives over R_Σ. Then $(R_\Sigma)_{\Sigma'}$ is a universal localisation of R at a suitable set of maps between f.g. projectives over R.

Proof: Let $\alpha : P' \to Q'$ be a map between stably induced f.g. projectives over R_Σ. Since P' and Q' are stably induced, there is an integer n such that $P' \oplus R^n$ and $Q' \oplus R^n$ are both induced modules. Clearly, α is stably associated to $\alpha \oplus I_n$, which is a map between induced modules and as such is stably associated to an induced map by 4.4. Consequently α is stably associated to an induced map. Adjoining a universal inverse to a map however has exactly the same effect as adjoining the universal inverse of a map stably associated to it, so we may replace all the elements of Σ' by a set $\bar{\Sigma}$ of induced maps that are stably associated to elements of Σ'. $(R_\Sigma)_{\Sigma'}$ is just $(R_\Sigma)_{\bar{\Sigma}}$ and this is clearly $R_{\Sigma \cup \bar{\Sigma}}$.

Homological properties of universal localisation

Let Σ be a set of maps between f.g. left projectives over R. It is clear that the homomorphism from R to R_Σ is an epimorphism in the category of rings, since there can be at most one inverse to a given map

$\alpha:P \to Q$ in the category of modules over a ring. Consequently, the category of right R_Σ modules may be regarded as a full subcategory of the category of right R modules. It turns out that this subcategory has some interesting properties that form the content of this section which were worked out with Warren Dicks. Our next result is a development of an argument due to Dlab and Ringel (84).

<u>Theorem 4.7</u> The category of right R_Σ modules is closed under extensions in the category of right R modules. Therefore, $\mathrm{Ext}^1_R(M,N) = \mathrm{Ext}^1_{R_\Sigma}(M,N)$ for R_Σ modules M and N.

Proof: We may characterise the R_Σ modules amongst the R modules by the property that for each $\alpha \in \Sigma$, $\alpha : P \to Q$, and for each R_Σ module M, the induced map $1\otimes_R\alpha:M\otimes_R P \to M\otimes_R Q$ is bijective. It is now clear that an extension in the category of R modules of a pair of R_Σ modules must be an R_Σ module. It follows trivially that $\mathrm{Ext}^1_R(M,N) = \mathrm{Ext}^1_{R_\Sigma}(M,N)$ for R_Σ modules M and N.

Our next result gives a number of consequences of the conclusion of this result.

<u>Theorem 4.8</u> Let $R \to S$ be an epimorphism in the category of rings; then the following conditions are equivalent:

a/ $\mathrm{Ext}^1_R = \mathrm{Ext}^1_S$ on left S modules;

b/ $\mathrm{Tor}^R_1(S,S) = 0$;

c/ $\mathrm{Tor}^R_1(M,N) = \mathrm{Tor}^S_1(M,N)$ for S modules M and N;

d/ $\mathrm{Ext}^1_R = \mathrm{Ext}^1_S$ on right S modules.

Proof: a $\Rightarrow$ B: let $0 \to M \to F \to S \to 0$ be a short exact sequence of right R modules where F is free. Then, we have the exact sequence:

1/ $0 \to \mathrm{Tor}^R_1(S,S) \to M\otimes_R S \to F\otimes_R S \to S\otimes_R S = S \to 0$;

we also have the push-out diagram:

2/
$$\begin{array}{ccccccccc}
0 & \to & M & \to & F & \to & S & \to & 0 \\
 & & \downarrow & & \downarrow & & \| & & \\
0 & \to & M\otimes_R S & \to & N & \to & S & \to & 0
\end{array}$$

by assumption, N must be an S module, and so, we may factor 2/ through 1/ in the following sense; we have a commutative diagram with exact rows:

$$\begin{array}{ccccccccccc}
 & & 0 & \to & M & \to & F & \to & S & \to & 0 \\
 & & & & \downarrow & & \downarrow & & \| & & \\
0 & \to & \mathrm{Tor}^R_1(S,S) & \to & M\otimes_R S & \to & F\otimes_R S & \to & S & \to & 0 \\
 & & & & \| & & \downarrow & & \| & & \\
 & & 0 & \to & M\otimes_R S & \to & N & \to & S & \to & 0
\end{array}$$

It follows that $Tor_1^R(S,S) = 0$.

$b \Rightarrow c$: let M be a right S module and let $0 \to A \to F \to M \to 0$ be an exact sequence of S modules where F is free; then we have the exact sequence:

$$0 = Tor_1^R(F,S) \to Tor_1^R(M,S) \to A\otimes_R S \to F\otimes_R S \to M\otimes_R S \to 0.$$

Since $R \to S$ is an epimorphism in the category of rings $B\otimes_R S \cong B$ for any S module, B, so $Tor_1^R(M,S) = 0$. Let N be a left S module, and let $0 \to C \to G \to N \to 0$ be an exact sequence of S modules where G is free; then

$$0 = Tor_1^R(M,G) \to Tor_1^R(M,N) \to M\otimes_R C \to M\otimes_R G \to M\otimes_R N \to 0$$

is an exact sequence. For any left S module D, $M\otimes_R D$ is isomorphic to $M\otimes_S D$ by the natural map since $R \to S$ is an epimorphism; therefore, the last exact sequence shows that $Tor_1^R(M,N) \cong Tor_1^S(M,N)$ for arbitrary S modules M and N.

$c \Rightarrow a$: let $0 \to M \to A \to N \to 0$ be an exact sequence of right R modules where M and N are S modules; then because $Tor_1^R(N,S) = 0$, we have the commutative diagram with exact rows:

$$\begin{array}{ccccccccc} 0 & \to & M & \to & A & \to & N & \to & 0 \\ & & \downarrow & & \downarrow & & \downarrow & & \\ 0 & \to & M\otimes_R S & \to & A\otimes_R S & \to & N\otimes_R S & \to & 0 \end{array}$$

$M \cong M\otimes_R S$ and $N \cong N\otimes_R S$, so by the 5-lemma $A \cong A\otimes_R S$, which shows that A must be an S module. It follows that $Ext_R^1 \cong Ext_S^1$ on right S modules.

We have shown that a,b, and c are equivalent; by symmetry, d is equivalent to them too.

In particular, a universal localisation satisfies all these conditions. It is clear that an epimorphism satisfying the conditions of theorem 4.8 need not be a universal localisation, since if I is an ideal of a ring R such that $I = I^2$, then $Tor_1^R(R/I,R/I) = 0$, but R/I is seldom a universal localisation of R.

The preceding results allow us to prove a theorem due to Dicks and Bergman.

Theorem 4.9 The universal localisation of a right hereditary ring is right hereditary.

Proof: A ring R is right hereditary if and only if Ext_R^1 is a right exact functor in the second variable. If R_Σ is a universal localisation of R, then $\mathrm{Ext}_{R_\Sigma}^1$ on R_Σ modules is isomorphic to Ext_R^1 which is right exact in the second variable. Therefore, R_Σ must be a right hereditary ring.

Universal localisation and ring coproducts

There is a useful way to obtain the ring coproduct of two rings amalgamating a common subring by considering a suitable universal localisation.

Theorem 4.10 Let A and B be R-rings; let $T = \begin{pmatrix} A & A \otimes_R B \\ 0 & B \end{pmatrix}$ and consider the map $\alpha : (0 \quad B) \to (A \quad A \otimes_R B)$ given by left multiplication by $\begin{pmatrix} 0 & 1 \otimes 1 \\ 0 & 0 \end{pmatrix}$; then $T_\alpha \cong M_2(A \sqcup_R B)$.

Proof: The elements $e_{11} = \begin{pmatrix} 1 & 0 \\ 0 & 0 \end{pmatrix}$, $e_{22} = \begin{pmatrix} 0 & 0 \\ 0 & 1 \end{pmatrix}$, α, and α^{-1} of T_α form a set of matrix units; the centraliser of these matrix units is isomorphic to $e_{11} T_\alpha e_{11}$, which is generated by a copy of A and a copy of B subject only to the relation that R is amalgamated.

This has the following consequence.

Theorem 4.11 Let S_1 and S_2 be semisimple artinian R-rings for some ring R: then $S_1 \sqcup_R S_2$ is hereditary.

Proof: $M_2(S_1 \sqcup_R S_2)$ is a universal localisation of the hereditary ring

$$\begin{pmatrix} S_1 & S_1 \otimes_R S_2 \\ 0 & S_2 \end{pmatrix}$$

Algebraic K-theory of universal localisation

We mentioned earlier that Ore localisation is a special case of universal localisation at sets of maps between f.g. projectives; in particular, central localisation is a special case. Bass and Murthy (67) found an exact sequence connecting the K-theory of R with the K-theory of R_S for a central subset S of R, from K_1 down to K_0, and later, Gersten (75) was able to generalise this to a long exact sequence for central localisation. Recently, Cohn (82) and Revesz (84) were able to calculate the K_1 of the free skew field and it was possible to rephrase this as Revesz did in terms of an exact sequence connecting K_0 and K_1 of the free algebra with the K-theory of the free skew field. Here we shall find a common generalisation of the Bass-Murthy result and the Cohn-Revesz theorem. The reason for proving such a result lies not just in allowing us to calculate K_1 of a universal localisation, but also in its giving us a way of quantifying Malcolmson's criterion.

First of all, we define a useful category; the definition of this category is very natural in the context of universal localisation though rather less obvious in the Ore case. Let R be a ring and Σ a collection of maps between f.g. projectives such that R embeds in R_Σ. Let $\bar{\Sigma}$ be a collection of maps between f.g. projectives that become invertible over R_Σ such that all maps between f.g. projectives over R that have inverses in R_Σ are associated to some map in $\bar{\Sigma}$; clearly, $R_{\bar{\Sigma}}$ is just R_Σ. Further, all elements of Σ are injective because of our assumption that R should embed in $R_{\bar{\Sigma}}$. Let $\underline{T}$ be the full subcategory of the category of f.p. modules over R whose objects are the cokernels of elements of $\bar{\Sigma}$. This category is independent of our choice of $\bar{\Sigma}$ and is closed under extensions in the category of modules over R, since $\bar{\Sigma}$ may be chosen to be lower multiplicatively closed.

<u>Theorem 4.12</u> Let R be a ring and Σ a collection of maps between f.g. projectives such that R embeds in R_Σ; let $\bar{\Sigma}$ and $\underline{T}$ be as we defined in the foregoing discussion. Then there is an exact sequence,

$$K_1(R) \overset{r}{\to} K_1(R_\Sigma) \overset{s}{\to} K_0(\underline{T}) \overset{t}{\to} K_0(R) \overset{u}{\to} K_0(R\ ) \quad ,$$

where r and u are induced by the ring homomorphism, t is given by the map $[M_\alpha] \to [Q] - [P]$, where $0 \to P \overset{\alpha}{\to} Q \to M_\alpha \to 0$ is an exact sequence, and α is an element of $\bar{\Sigma}$, and s will be defined in the course of the proof.

Proof: Throughout this proof, M_β will denote the cokernel of β.

Let $\alpha:P \to Q$ be an isomorphism between induced f.g. projectives over R_Σ; by Cramer's rule, we have an equation:

$$\beta\begin{pmatrix} I & \alpha_1 \\ O & \alpha \end{pmatrix} = \beta'$$

where β and β' are defined over R and α_1 is defined over R_Σ. Since β and $\begin{pmatrix} I & \alpha_1 \\ O & \alpha \end{pmatrix}$ are invertible over R_Σ, so is β', so it is associated to an element of $\bar{\Sigma}$. We attempt to define a map from such an invertible map to $K_O(\underline{T})$ by $s(\alpha) = [M_{\beta'}] - [M_\beta]$. We need to show that this is well-defined and its restriction to automorphisms defines a homomorphism from $K_1(R_\Sigma)$ to $K_O(\underline{T})$. Assuming that we have proven this, it is not too hard to complete the proof of the theorem, so we shall demonstrate this first.

<u>1</u> Exactness at $K_O(R)$

Certainly, $tu = O$; for given $M_\alpha \in \underline{T}$, $t([M_\alpha]) = [Q] - [P]$, where M_α has a presentation $O \to P \overset{\alpha}{\to} Q \to M_\alpha \to O$; since α becomes an isomorphism over R_Σ, $u(Q) - u(P) = O$.

Conversely, if $u([P_2] - [P_1]) = O$, we have an equation over R_Σ, $R_\Sigma \otimes_R P_1 \oplus R_\Sigma^n \cong R_\Sigma \otimes_R P_2 \oplus R_\Sigma^n$ for some n; so there is an isomorphism between induced f.g. projectives $\alpha:R_\Sigma \otimes_R (P_1 \oplus R^n) \to R_\Sigma \otimes_R (P_2 \oplus R^n)$, and so, by Cramer's rule, we have an equation:

$$\beta\begin{pmatrix} I_Q & \alpha_1 \\ O & \alpha \end{pmatrix} = \beta'$$

for maps defined over R, $\beta:Q' \to Q \oplus P_1 \oplus R^n$, $\beta':Q' \to Q \oplus P_2 \oplus R^n$, where β,β' are associated to elements of $\bar{\Sigma}$ (since α is invertible over R_Σ), α_1 is some map defined over R_Σ, and I_Q is the identity map on Q.

Hence, $t([M_{\beta'}] - [M_\beta])$
$= [Q] + [P_2] + n - [Q'] - [Q] - [P_1] - n + [Q']$, which is $[P_2] - [P_1]$, so that we have exactness at this point.

<u>2</u> Exactness at $K_O(\underline{T})$

Let $\alpha:P \to P$ be an isomorphism on an induced f.g. projective. Then we defined $s(\alpha) = [M_{\beta'}] - [M_\beta]$, where we have an equation:

$$\beta \begin{pmatrix} I_Q & \alpha_1 \\ 0 & \alpha \end{pmatrix} \beta'$$

where β and $\beta':Q' \to Q \oplus P$ are both associated to maps in $\bar{\Sigma}$. So we find that

$$st(\alpha) = [Q] + [P] - [Q'] - [Q] - [P] + [Q'] = 0 .$$

Conversely, suppose that $t([M_{\alpha_1}] - [M_{\alpha_2}]) = 0$; then if we have presentations $0 \to P_i \overset{\alpha_i}{\to} Q_i \to M_i \to 0$ for $i = 1,2$, $[Q_1] - [P_1] = [Q_2] - [P_2]$, so for suitable n, we have an equation, $Q_1 \oplus P_2 \oplus R^n \cong Q_2 \oplus P_1 \oplus R^n$. We construct from this, presentations for M_1 and M_2 $0 \to P_i \overset{\beta_i}{\to} Q_i \to M_i \to 0$ for suitable P and Q. This is easy for

$0 \to P_1 \oplus P_2 \oplus R^n \overset{\alpha_1 \oplus I \oplus I}{\to} Q_1 \oplus P_2 \oplus R^n \to M_1 \to 0$ is a presentation for M_1,

and $0 \to P_1 \oplus P_2 \oplus R^n \overset{I \oplus \alpha_2 \oplus I}{\to} P_1 \oplus Q_2 \oplus R^n \to M_2 \to 0$ is a presentation for M_2.

Consider the map over R_Σ defined by $\beta_2^{-1}\beta_1 : R_\Sigma \otimes_R Q \to R_\Sigma \otimes_R Q$. Because $\beta_2(\beta_2^{-1}\beta_1) = \beta_1$, we see that $s(\beta_2^{-1}\beta_1) = [M_{\beta_1}] - [M_{\beta_2}] = [M_{\alpha_1}] - [M_{\alpha_2}]$. So we have shown exactness at $K_0(\underline{T})$.

<u>3</u> Exactness at $K_1(R_\Sigma)$

This is rather harder to show than the previous two parts.

Certainly, $rs = 0$; for, if $\alpha:P \to P$ is an automorphism over R, the equation $1.\alpha = \alpha$ satisfies the condition of Cramer's rule, so $rs(\alpha) = [M_\alpha] - [M_\alpha] = 0$.

Conversely, suppose that $\alpha:P \to P$ is an automorphism of an induced f.g. projective over R_Σ such that $s(\alpha) = 0$; we have an equation:

$$\beta \begin{pmatrix} I_Q & \alpha_1 \\ 0 & \alpha \end{pmatrix} = \beta'$$

where $\beta:Q' \to Q \oplus P$, $\beta':Q' \to Q \oplus P$ are associated to elements of $\bar{\Sigma}$ and we deduce that $0 = s(\alpha) = [M_{\beta'}] - [M_\beta]$.

We wish to show that the class of α in $K_1(R_\Sigma)$ lies in the image of $K_1(R)$ in $K_1(R_\Sigma)$, and our equation shows that the class of α in $K_1(R_\Sigma)$ is equal to that of $\beta^{-1}\beta'$; so we need to examine what it means if $[M_\beta] = [M_{\beta'}]$ in $K_0(\underline{T})$.

We define a recursive relation $\sim$ on the objects of an exact category $\underline{E}$, that amounts to the relation $[X] = [Y]$ in $K_0(\underline{E})$.

We start our recursive relation by defining that $X \sim Y$, if $X \cong Y$; next, if $X \sim X'$, and $Z \sim Z'$ and we have exact sequences $0 \to X \to Y \to Z \to 0$, and $0 \to X' \to Y' \to Z' \to 0$, then $Y \sim Y'$; similarly, if we have the first exact sequence, and in place of the second we have the exact sequence $0 \to Z' \to Y' \to X' \to 0$ we extend the relation to $Y \sim Y'$.

Next, if $X \sim X'$, and $Y \sim Y'$, and we have exact sequences

$0 \to X \to Y \to Z \to 0$

$0 \to X' \to Y' \to Z' \to 0$, then $Z \sim Z'$; and if we have exact sequences

$0 \to X \to Y \to Z \to 0$

$0 \to Z' \to Y' \to X' \to 0$, then $Z \sim Z'$.

We repeat these operations as often as they apply, and finally we make sure that the relation is transitive by defining that if $X \sim X'$ and $X' \sim X''$, then $X \sim X''$.

We examine what this means in the category $\underline{T}$; that is, we attempt to find the equivalence relation defined on the maps by $M_\alpha \sim M_\beta$ for α, β in $\bar{\Sigma}$.

$M_\alpha \cong M_\beta$ if and only if α is stably associated to β. Our first operation for generating the equivalence relation on the modules induces the relation on maps that if $M_\alpha \sim M_{\alpha'}$, and $M_\gamma \sim M_{\gamma'}$ then $M_{\begin{pmatrix}\alpha & \\ 0 & \gamma\end{pmatrix}} \sim M_{\begin{pmatrix}\alpha' & \\ 0 & \gamma'\end{pmatrix}}$ and $M_{\begin{pmatrix}\alpha & \\ 0 & \gamma\end{pmatrix}} \sim M_{\begin{pmatrix}\alpha' & 0 \\ & \gamma'\end{pmatrix}}$. Our second operation for generating the equivalence on modules induces the relation on maps that if $M_\alpha \sim M_{\alpha'}$, and $M_{\begin{pmatrix}\alpha & \\ 0 & \gamma\end{pmatrix}} \sim M_{\begin{pmatrix}\alpha' & \\ 0 & \gamma'\end{pmatrix}}$ or $M_{\begin{pmatrix}\alpha & \\ 0 & \gamma\end{pmatrix}} \sim M_{\begin{pmatrix}\alpha' & 0 \\ & \gamma'\end{pmatrix}}$ then $M_\gamma \sim M_{\gamma'}$.

We have that $M_\beta \sim M_{\beta'}$ for the maps β and $\beta' : Q' \to Q \oplus P$, so we know that we can pass along some chain of simple operations of the type described above to show this equivalence; by passing to $\begin{pmatrix}\beta & 0 \\ 0 & I_m\end{pmatrix}$ and $\begin{pmatrix}\beta' & 0 \\ 0 & I_n\end{pmatrix}$ for suitably large m and n, we may assume that any stable associations that occur in this chain are actually associations. We wish to show that this equivalence on maps forces the image of $\beta'^{-1}\beta$ in $K_1(R_\Sigma)$ to lie in the image of $K_1(R)$ in $K_1(R_\Sigma)$.

If $\alpha, \alpha' : P_1 \to Q_1$ are associated over R, the class of the map $\alpha^{-1}\alpha' : Q_1 \to Q_1$ over R_Σ in $K_1(R_\Sigma)$ clearly lies in the image of $K_1(R)$.

If $\alpha_i, \alpha_i' : P_i \to Q_i$ for $i = 1,2$ satisfy $\alpha_i^{-1}\alpha_i' : Q_i \to Q_i$ lies in the image of $K_1(R)$ in $K_1(R_\Sigma)$, then from the equations

$$\begin{pmatrix} \alpha_1 & \beta \\ 0 & \alpha_2 \end{pmatrix} = \begin{pmatrix} I & 0 \\ 0 & \alpha_2 \end{pmatrix}\begin{pmatrix} I & \beta \\ 0 & I \end{pmatrix}\begin{pmatrix} \alpha_1 & 0 \\ 0 & I \end{pmatrix}, \quad \begin{pmatrix} \alpha_1 & 0 \\ \gamma & \alpha_2 \end{pmatrix} = \begin{pmatrix} \alpha_1 & 0 \\ 0 & I \end{pmatrix}\begin{pmatrix} I & 0 \\ \gamma & I \end{pmatrix}\begin{pmatrix} I & 0 \\ 0 & \alpha_2 \end{pmatrix}$$

and since the class of $\begin{pmatrix} I & \beta \\ 0 & I \end{pmatrix}$ and $\begin{pmatrix} I & 0 \\ \gamma & 0 \end{pmatrix}$ are trivial in $K_1(R_\Sigma)$, we see that the image of $\begin{pmatrix} \alpha_1 & \beta \\ 0 & \alpha_2 \end{pmatrix}$ in $K_1(R_\Sigma)$ is equal to the image of $\begin{pmatrix} \alpha_1 & 0 \\ 0 & \alpha_2 \end{pmatrix}$ and the image of $\begin{pmatrix} \alpha_1 & 0 \\ \gamma & \alpha_2 \end{pmatrix}$. So we need to check that the image of $\begin{pmatrix} \alpha_1 & 0 \\ 0 & \alpha_2 \end{pmatrix}^{-1}\begin{pmatrix} \alpha_1' & 0 \\ 0 & \alpha_2' \end{pmatrix}$ in $K_1(R_\Sigma)$ lies in the image of $K_1(R)$, which is clear.

Finally, if $\begin{pmatrix} \alpha_1 & 0 \\ 0 & \alpha_2 \end{pmatrix}^{-1}\begin{pmatrix} \alpha_1' & 0 \\ 0 & \alpha_2' \end{pmatrix}$ lies in the image of $K_1(R)$ in $K_1(R_\Sigma)$ and $\alpha_1^{-1}\alpha_1'$ does too, then so must $\alpha_2^{-1}\alpha_2'$, which shows that the second operation generating the equivalence relation $M_\gamma \sim M_\gamma'$ preserves the property that $\gamma^{-1}\gamma'$ lies in $K_1(R)$.

We are left with showing that our map s from isomorphisms of induced f.g. projectives is well-defined and induces a map from $K_1(R_\Sigma)$ to $K_0(\underline{T})$. This is where Malcolmson's criterion is most useful.

Suppose that $\beta : R_\Sigma \otimes_R P \to R_\Sigma \otimes_R Q$ is an isomorphism between induced f.g. projectives and that we have two equations defined by Cramer's rule:

$$(\alpha|\alpha_1)\left(\begin{array}{c|c} I & \beta_1 \\ \hline 0 & \beta \end{array}\right) = (\alpha|\alpha') \quad , \quad (\gamma|\gamma_1)\left(\begin{array}{c|c} I & \beta_2 \\ \hline 0 & \beta \end{array}\right) = (\gamma|\gamma')$$

and $(\alpha|\alpha_1)$, $(\alpha|\alpha')$, $(\gamma|\gamma_1)$ and $(\gamma|\gamma')$ are associated to elements of $\bar{\Sigma}$; then we know that

$$\begin{pmatrix} \alpha & \alpha_1 & 0 & | & 0 \\ 0 & \gamma_1 & \gamma & | & \gamma_1 \end{pmatrix} \left(\begin{array}{c|c} I & \begin{matrix}\beta_1 \\ \beta^1 \\ \beta_2\end{matrix} \\ \hline 0 & \beta-\beta \end{array}\right) = \begin{pmatrix} \alpha & \alpha & 0 & | & \alpha' \\ 0 & \gamma_1 & \gamma & | & \gamma' \end{pmatrix}$$

so, $(0 \;\; 0 \;\; 0 \mid I) \begin{matrix} \alpha & \alpha_1 & 0 & 0 \\ 0 & \gamma_1 & \gamma & \gamma_1 \end{matrix}^{-1} \begin{matrix} \alpha' \\ \gamma' \end{matrix} = 0$;

by Malcolmson's criterion, we have an equation:

$$\left(\begin{array}{cccccc|c} \alpha & \alpha_1 & 0 & 0 & 0 & 0 & \alpha' \\ 0 & \gamma_1 & \gamma & \gamma_1 & 0 & 0 & \gamma' \\ 0 & 0 & 0 & 0 & \delta_1 & 0 & 0 \\ 0 & 0 & 0 & 0 & 0 & \delta_2 & \varepsilon_2 \\ \hline 0 & 0 & 0 & I & \varepsilon_1 & 0 & 0 \end{array}\right) = \left(\begin{array}{c} \mu_1 \\ \mu_2 \\ \mu_3 \\ \mu_4 \\ \hline \emptyset \end{array}\right) (\nu \mid \tau)$$

where $\delta_1, \delta_2, \begin{pmatrix}\mu_1 \\ \mu_2 \\ \mu_3 \\ \mu_4\end{pmatrix}$ and τ are associated to elements of $\bar{\Sigma}$.

From this, we construct two equations:

$$1 \quad \left(\begin{array}{cc} \mu_1 & 0 \\ \mu_2 & \gamma' \\ \mu_3 & 0 \\ \mu_4 & 0 \\ \hline \emptyset & 0 \end{array}\right) \left(\begin{array}{c|c} \nu & 0 \\ \hline 0 & I \end{array}\right) = \left(\begin{array}{cccccc|c} \alpha & \alpha_1 & 0 & 0 & 0 & 0 & 0 \\ 0 & \gamma_1 & \gamma & \gamma_1 & 0 & 0 & \gamma' \\ 0 & 0 & 0 & 0 & \delta_1 & 0 & 0 \\ 0 & 0 & 0 & 0 & 0 & \delta_2 & 0 \\ \hline 0 & 0 & 0 & I & \varepsilon_1 & 0 & 0 \end{array}\right)$$

and also

$$2 \quad \left(\begin{array}{cc} \mu_1 & 0 \\ \mu_2 & \gamma' \\ \mu_3 & 0 \\ \mu_4 & 0 \\ \hline \emptyset & 0 \end{array}\right) \left(\begin{array}{c|c} \nu & -\tau \\ \hline 0 & I \end{array}\right) = \left(\begin{array}{cccccc|c} \alpha & \alpha_1 & 0 & 0 & 0 & 0 & -\alpha' \\ 0 & \gamma_1 & \gamma & \gamma_1 & 0 & 0 & 0 \\ 0 & 0 & 0 & 0 & \delta_1 & 0 & 0 \\ 0 & 0 & 0 & 0 & 0 & \delta_2 & -\varepsilon_2 \\ \hline 0 & 0 & 0 & I & \varepsilon_1 & 0 & 0 \end{array}\right)$$

First of all, it is clear that RHS(1) and RHS(2) lie in $\bar{\Sigma}$, and $[M_{RHS(1)}] = [M_{\delta_1}] + [M_{\delta_2}] + [M_{(\alpha\alpha_1)}] + [M_{(\gamma\gamma')}]$, whilst $[M_{RHS(2)}] = [M_{\delta_1}] + [M_{\delta_2}] + [M_{(\alpha\alpha')}] + [M_{(\gamma\gamma_1)}]$.

If $X,Y \in \bar{\Sigma}$, $[M_{XY}] = [M_X] + [M_Y]$ when XY is defined. Since

$\begin{pmatrix} \nu & 0 \\ 0 & 1 \end{pmatrix}$ lies in $\bar{\Sigma}$,

$\begin{pmatrix} \mu_1 & 0 \\ \mu_2 & \gamma' \\ \mu_3 & 0 \\ \mu_4 & 0 \\ \emptyset & 0 \end{pmatrix} = \mu$ must be associated to an element in $\bar{\Sigma}$ and

$$[M_{LHS(1)}] = [M_\mu] + [M_\nu] = [M_\mu] \quad [M_{\begin{smallmatrix} \nu & -\tau \\ 0 & I \end{smallmatrix}}] = [M_{LHS(2)}]$$

We deduce that $[M_{(\alpha\alpha_1)}] + [M_{(\gamma\gamma')}] = [M_{(\alpha\alpha')}] + [M_{(\gamma\gamma_1)}]$, which shows that s is well-defined.

We need to show that it induces a homomorphism from $K_1(R_\Sigma)$ to $K_0(\underline{T})$, and to do this, we need to show that $s\begin{pmatrix} \beta & 0 \\ 0 & I \end{pmatrix} = s(\beta)$, and $s(\beta\gamma) = s(\beta) + s(\gamma)$. It is convenient to show that $s\begin{smallmatrix} I\beta_1 \\ 0\beta \end{smallmatrix} = s(\beta)$ for arbitrary β_1.

Suppose that we have equations:

$$(\alpha|\alpha_1)\left(\begin{array}{c} I \;\; \beta_2 \\ \hline 0 \;\; \beta \end{array}\right) = (\alpha|\alpha')\quad , \quad (\gamma|\gamma_1)\left(\begin{array}{c|c} I & \beta_3 \\ \hline 0 & \beta_1 \end{array}\right) = (\gamma|\gamma')$$

then we find that

$$\left(\begin{array}{cc|cc} \gamma & \gamma_1 & 0 & 0 \\ \hline 0 & 0 & \alpha & \alpha_1 \end{array}\right)\left(\begin{array}{c|cc} I & 0 & \beta_3 \\ & & \beta_1 \\ \hline 0 & I & \beta_2 \\ 0 & 0 & \beta \end{array}\right) = \left(\begin{array}{cc|cc} \gamma & \gamma_1 & 0 & \gamma' \\ \hline 0 & 0 & \alpha & \alpha' \end{array}\right)$$

operating by a permutation matrix internally on the left hand side of this equation does not alter $[M_{\begin{pmatrix} \gamma\gamma_1 00 \\ 00\alpha\alpha_1 \end{pmatrix}}]$ in any way and allow us to change the second matrix on the left hand side to the form $\begin{pmatrix} I & 0 & \beta_3 \\ & 0 & \beta_2 \\ 0 & I & \beta_1 \\ 0 & 0 & \beta \end{pmatrix}$; so

$$s\begin{pmatrix} I & \beta_1 \\ 0 & \beta \end{pmatrix} = [M_{\begin{pmatrix} \gamma & \gamma_1 & 0 & \gamma' \\ 0 & 0 & \alpha & \alpha' \end{pmatrix}}] - [M_{\begin{pmatrix} \gamma & \gamma_1 & 0 & 0 \\ 0 & 0 & \alpha & \alpha_1 \end{pmatrix}}] = [M_{(\alpha\alpha')}] - [M_{(\alpha\alpha_1)}] = \delta(\beta)$$

as we wished to show.

In order to show that $s(\beta\gamma) = s(\beta) + s(\gamma)$, we first show it under the assumption that $\beta \in \bar{\Sigma}$.

$s\begin{pmatrix} I & \gamma \\ O & \beta\gamma \end{pmatrix} = s(\beta\gamma)$ by our last step; it is clear that $s(\delta\varepsilon) = s(\varepsilon)$ if δ is an isomorphism over R; so we find that

$$s(\beta\gamma) = s\begin{pmatrix} I & \gamma \\ O & \beta\gamma \end{pmatrix} = s\begin{pmatrix} \beta & O \\ I & \gamma \end{pmatrix}$$

If $(\alpha|\alpha_1)\left(\begin{array}{c|c} I & \gamma_1 \\ \hline O & \gamma \end{array}\right) = (\alpha|\alpha')$, $\left(\begin{array}{c|cc} I & O & O \\ \hline O & \alpha & \alpha_1 \end{array}\right)\left(\begin{array}{c|cc} \beta & O & O \\ \hline O & I & \gamma_1 \\ I & O & \gamma \end{array}\right) = \left(\begin{array}{c|c} \beta & O \\ \hline \alpha_1 & \alpha\alpha' \end{array}\right)$

so $s\begin{pmatrix} \beta & O \\ I & \gamma \end{pmatrix} = [M_\beta] + [M_{(\alpha\alpha')}] - [M_{(\alpha\alpha_1)}] = s(\beta) + s(\gamma)$, and

$s(\beta\gamma) = s\begin{pmatrix} \beta & O \\ I & \gamma \end{pmatrix} = s(\beta) + s(\gamma)$.

Finally, we deal with the general case.

Assume $(\delta|\delta_1)\left(\begin{array}{c|c} I & \beta_1 \\ \hline O & \beta \end{array}\right) = (\delta|\delta')$ then, by the last step,

$$s(\delta|\delta_1) + s\left(\begin{array}{c|c} I & O \\ \hline O & \beta\gamma \end{array}\right) = s\ (\delta|\delta_1)\left(\begin{array}{c|c} I & O \\ \hline O & \beta\gamma \end{array}\right) = s\left((\delta|\delta')\begin{pmatrix} I & -\delta_1\gamma \\ O & \gamma \end{pmatrix}\right) = s(\delta|\delta') + s(\gamma)$$

so $s(\beta\gamma) = s\begin{pmatrix} I & O \\ O & \beta\gamma \end{pmatrix} = s(\gamma) + s(\delta\delta') - s(\delta\delta_1) = s(\gamma) + s(\beta)$ which shows that s induces a homomorphism from $K_1(R_\Sigma)$ to $K_O(\underline{T})$, and so completes the proof of our theorem.

The Bass-Murthy sequence is a special case of this result; it would be of interest to know whether this sequence may be extended to a long exact sequence of K-groups for universal localisation as Gersten does for central localisation. It appears very likely for special cases such as the passage from a fir to its universal skew field of fractions. This leads to the following conjecture for $K_2(F)$, where F is the free skew field on some set X over an algebraically closed field, k. $K_2(F)$ should be $K_2(k) \times \underset{A}{X} k^\times$, where A is the set of stable association classes of full matrix atoms over $k\langle X\rangle$.

5 UNIVERSAL HOMOMORPHISMS FROM HEREDITARY TO SIMPLE ARTINIAN RINGS

In this chapter, we shall prove our main theorems on universal homomorphisms from hereditary rings to simple artinian rings; we shall find that these arise as universal localisations at the set of maps between f.g. projective modules that are full with respect to a projective rank function that takes values in $\frac{1}{n}\mathbb{Z}$. In doing so, we shall also develop techniques for studying intermediate localisations, which generalise the results known to hold for firs, and which will lead in later chapters to a detailed analysis of the subring structure of these universal localisations of hereditary rings. However, the most interesting result of this chapter in the short term is the construction of the simple artinian coproduct with amalgamation. This is given by constructing a universal homomorphism from the ring $S_1 \underset{S_0}{\sqcup} S_2$ to a simple artinian ring $S_1 \underset{S_0}{\sqcup} S_2$, where S_i is simple artinian. In order to assist the reader in understanding the method, we give a brief outline of what we shall do in this case.

$S_1 \underset{S_0}{\sqcup} S_2$ is a hereditary ring with a unique projective rank function ρ by 3.1. We consider what maps between f.g. projectives have a chance of becoming invertible under a suitable homomorphism to a simple artinian ring; firstly, the homomorphism must induce the projective rank function ρ on $S_1 \underset{S_0}{\sqcup} S_2$ since this is the only rank function. Therefore, if $\alpha: P \to Q$ is a map that becomes invertible under a homomorphism to a simple artinian ring, $\rho(P) = \rho(Q)$, and α cannot factor through a projective of smaller rank; that is, α is a full map with respect to the rank function. At this point, we consider the ring obtained by adjoining the universal inverses of all full maps with respect to the rank function, since Cohn (71) has shown that in the case where S_i are all skew fields that the result is a skew field; we are able to show in this case that this universal localisation is simple artinian as we wanted it to be. We have been calling such constructions by such names as the 'simple artinian coproduct' or the 'universal homomorphism from the hereditary ring R to a simple artinian ring inducing the projective rank function ρ; we shall not justify these names immediately, but in chapter 7 when we discuss homomorphisms to simple artinian rings in greater generality, we shall see that all other homomorphisms from hereditary rings to simple artinian rings inducing the given projective rank function are specialisations of the one referred to as universal.

In fact, the general method applies to constructing universal homomorphisms from rings with a Sylvester projective rank function to von Neumann regular rings, so we shall deal with this generality in the opening stages of this chapter; we shall complete the discussion of homomorphisms to von Neumann regular rings in the next chapter.

Universal localisation at a Sylvester projective rank function

Theorem 5.1 Let R be a ring with a Sylvester projective rank function ρ having enough right and left full maps (see 1.16, and the following definition). Let Σ be a collection of full maps with respect to ρ between f.g projectives: then the projective rank function ρ extends to a Sylvester projective rank function on R, ρ_Σ, that has the same image as ρ and has enough right full and left full maps. The kernel of the map from R to R_Σ lies in the trace ideal of the projectives of rank 0.

Proof: First of all, we may assume that Σ is lower and upper multiplicatively closed (see the discussion before 4.6), since all elements of the lower and upper multiplicative closure of Σ are full with respect to ρ by 1.15 and become invertible in R_Σ. The use of this observation is that it puts us in good shape for both Cramer's rule and Malcolmson's criterion.

We prove the last statement of the theorem first, since it is a simple step.

Let $r \in R$ be in the kernel of $R \to R_\Sigma$; then, by Malcolmson's criterion, we have an equation:

$$1 \left(\begin{array}{cccc|c} 1 & 0 & 0 & 0 & r \\ 0 & 1 & 0 & 0 & 0 \\ 0 & 0 & \alpha_1 & 0 & 0 \\ 0 & 0 & 0 & \alpha_2 & \beta_2 \\ \hline 1 & 0 & \beta_1 & 0 & 0 \end{array}\right) = \frac{\delta_1}{\phi} (\delta_2 | \mu)$$

where $\alpha_i, \delta_i \in \Sigma$ and $\beta_i, \gamma, \phi, \mu$ are maps defined over R.

This shows that the nullity of LHS(1) is 1; by 1.15, and the fact that $1, \alpha_1, \alpha_2$ are full with respect to ρ, we may deduce that the nullity of $\begin{pmatrix} 1 & r \\ 1 & 0 \end{pmatrix}$ is 1, so that the nullity of r is 1, or that its inner projective rank with respect to ρ is 0, which implies that r must lie in the trace ideal of the projectives of rank 0 with respect to ρ. So, it is clear that R_Σ is not the zero ring.

Cramer's rule and Malcolmson's criterion give representations of maps over R_Σ, and ways of characterising when two representations are the same. We use these in order to define the rank of a map between induced f.g. projectives over R_Σ, and then to show that it is well-defined; then we show that this rank function on maps between induced projectives is the inner projective rank associated to a projective rank function that satisfies the law of nullity for these maps; it is not a hard step to show that the rank function must actually satisfy the law of nullity for all maps between f.g. projectives.

Given a map, $\beta : R_\Sigma \otimes_R P_1 \to R_\Sigma \otimes_R P_2$ between induced f.g. projectives, Cramer's rule gives us an equation:

$$2 \qquad (\alpha|\alpha_1)\left(\begin{array}{c|c} I_Q & \beta_1 \\ \hline 0 & \beta \end{array}\right) = (\alpha|\alpha')$$

where $(\alpha\alpha_1) \in \Sigma$, Q is a f.g. projective over R, β_1 is defined over R_Σ, and $(\alpha\alpha')$ is defined over R.

It is clear that if ρ extends to R_Σ, we should define
$\rho_\Sigma(\beta) = \rho(\alpha\alpha') - \rho(\alpha) = \rho(\alpha\alpha') - \rho(Q) = \rho(\alpha\alpha') - \rho(\alpha\alpha_i) + \rho(P_1)$

The first definition shows that $\rho(\beta)$ is non-negative, if it is well-defined. Suppose that we have a second equation:

$$3 \qquad (\gamma|\gamma_1)\left(\begin{array}{c|c} I_{Q'} & \beta_2 \\ \hline 0 & \beta \end{array}\right) = (\gamma|\gamma')$$

where $(\gamma\gamma_1) \in \Sigma$, Q' is a f.g. projective over R, β_2 is defined over R_Σ, and $(\gamma\gamma')$ is defined over R. Then, from (2) and (3) we construct the equation:

$$4 \qquad \left(\begin{array}{ccc|c} \alpha & \alpha_1 & 0 & 0 \\ 0 & \gamma_1 & \gamma & \gamma_1 \end{array}\right)\left(\begin{array}{c|c} & \beta_1 \\ I & \beta \\ & \beta_2 \\ \hline 0 & \beta-\beta \end{array}\right) = \left(\begin{array}{ccc|c} \alpha & \alpha_1 & 0 & \alpha' \\ 0 & \gamma_1 & \gamma & \gamma' \end{array}\right)$$

where I is an identity map on a suitable f.g. projective over R.

Over R_Σ, $0 = (0\ 0\ 0 \mid I) \begin{pmatrix} \alpha & \alpha_1 & 0 & 0 \\ 0 & \gamma_1 & \gamma & \gamma_1 \end{pmatrix}^{-1} \begin{pmatrix} \alpha' \\ \gamma' \end{pmatrix}$

By Malcolmson's criterion, we have an equation:

$$5 \qquad \left(\begin{array}{cccccc|c} \alpha & \alpha_1 & 0 & 0 & 0 & 0 & \alpha' \\ 0 & \gamma_1 & \gamma & \gamma_1 & 0 & 0 & \gamma' \\ 0 & 0 & 0 & 0 & \delta_1 & 0 & 0 \\ 0 & 0 & 0 & 0 & 0 & \delta_2 & \varepsilon_2 \\ \hline 0 & 0 & 0 & I & \varepsilon_1 & 0 & 0 \end{array}\right) = \begin{pmatrix} \mu \\ \hline \phi \end{pmatrix} (\nu \mid \tau)$$

where $\delta_i, \mu, \nu \in \Sigma$

We see that the right nullity of LHS(5) is $\rho(P_2)$. Since δ_1 and δ_2 are full, we see from 1.15, that the right nullity of $\begin{pmatrix} \alpha & \alpha_1 & 0 & 0 & \alpha' \\ 0 & \gamma_1 & \gamma & \gamma_1 & \gamma' \\ 0 & 0 & 0 & I & 0 \end{pmatrix}$ is $\rho(P_2)$. In turn, we look at the bottom row of this matrix of maps and apply 1.15 to deduce that the right nullity of $\begin{pmatrix} \alpha & \alpha_1 & 0 & \alpha' \\ 0 & \gamma_1 & \gamma & \gamma' \end{pmatrix}$ is $\rho(P_2)$. Since P_2 is the codomain of $\begin{pmatrix} \alpha' \\ \gamma' \end{pmatrix}$ and $\begin{pmatrix} \alpha & \alpha_1 & 0 \\ 0 & \gamma_1 & \gamma \end{pmatrix}$ is clearly right full (because $\begin{pmatrix} \alpha & \alpha_1 & 0 & 0 \\ 0 & \gamma_1 & \gamma & \gamma_1 \end{pmatrix}$ is full), we may write a minimal factorisation of $\begin{pmatrix} \alpha & \alpha_1 & 0 & \alpha' \\ 0 & \gamma_1 & \gamma & \gamma' \end{pmatrix}$ as:

$$6 \qquad \left(\begin{array}{ccc|c} \alpha & \alpha_1 & 0 & \alpha' \\ 0 & \gamma_1 & \gamma & \gamma' \end{array}\right) = \begin{pmatrix} \tau_1 \\ \tau_2 \end{pmatrix} (\chi_1 \mid \chi_2)$$

where $\rho \text{ codomain} \begin{pmatrix} \tau_1 \\ \tau_2 \end{pmatrix} = \rho \begin{pmatrix} \alpha & \alpha_1 & 0 \\ 0 & \gamma_1 & \gamma \end{pmatrix} = \rho \left(\text{codomain} \begin{pmatrix} \alpha & \alpha_1 & 0 \\ 0 & \gamma_1 & \gamma \end{pmatrix} \right)$

and so χ_1 must be a full map. From (6), we construct two equations

$$7 \qquad \begin{pmatrix} \tau_1 & 0 \\ \tau_2 & \gamma' \end{pmatrix} \left(\begin{array}{c|c} \chi_1 & 0 \\ 0 & I \end{array}\right) = \left(\begin{array}{ccc|c} \alpha & \alpha_1 & 0 & 0 \\ 0 & \gamma_1 & \gamma & \gamma' \end{array}\right)$$

$$8 \qquad \begin{pmatrix} \tau_1 & 0 \\ \tau_2 & \gamma' \end{pmatrix} \begin{pmatrix} \chi_1 & -\chi_2 \\ 0 & I \end{pmatrix} = \left(\begin{array}{ccc|c} \alpha & \alpha_1 & 0 & -\alpha' \\ 0 & \gamma_1 & \gamma & 0 \end{array}\right)$$

Using lemmas 1.15 and 1.13, we see that

$$\rho(\alpha\alpha_1) + \rho(\gamma\gamma') = \rho \begin{pmatrix} \tau_1 & 0 \\ \tau_2 & \gamma' \end{pmatrix} = \rho(\alpha\alpha') + \rho(\gamma\gamma_1)$$

Therefore, $\rho(\gamma\gamma') - \rho(\gamma\gamma_1) + \rho(P_1) = \rho(\alpha\alpha') - \rho(\alpha\alpha_1) + \rho(P_1)$ which shows that our definition of ρ_Σ is well-defined.

It is useful to show next that $\rho_\Sigma\begin{pmatrix}\alpha & O\\ O & \beta\end{pmatrix} = \rho_\Sigma(\alpha) + \rho_\Sigma(\beta)$ and that when $\alpha\beta$ is defined $\rho_\Sigma(\alpha\beta) \le \rho_\Sigma(\alpha), \rho_\Sigma(\beta)$.

So suppose that $(\gamma\gamma_1)\begin{pmatrix}I & \alpha_1\\ O & \alpha\end{pmatrix} = (\gamma\gamma')$ and $(\delta\delta_1)\begin{pmatrix}I & \beta_1\\ O & \beta\end{pmatrix} = (\delta\delta')$ with the usual conditions obtained from Cramer's rule. Then, if $\alpha\beta$ is defined we construct the equation:

$$\left(\begin{array}{ccc|c}\delta & \delta_1 & O & O\\ O & -\gamma' & \gamma & \gamma_1\end{array}\right)\left(\begin{array}{c|c} & \beta_1\\ I & \beta\\ & \alpha_1\beta\\ \hline O & \alpha\beta\end{array}\right) = \left(\begin{array}{ccc|c}\delta & \delta_1 & O & \delta'\\ O & -\gamma' & \gamma & O\end{array}\right)$$

so $\rho_\Sigma(\alpha\beta) = \rho\begin{pmatrix}\delta & \delta_1 & O & \delta'\\ O & -\gamma' & \gamma & O\end{pmatrix} - \rho\begin{pmatrix}\delta & \delta_1 & O & O\\ O & -\gamma' & \gamma & \gamma_1\end{pmatrix} + \rho(\text{dom }\alpha)$;

$\rho\begin{pmatrix}\delta & \delta_1 & O & \delta'\\ O & -\gamma' & \gamma & O\end{pmatrix} = \rho\begin{pmatrix}\delta & \delta' & O & -\delta_1\\ O & O & \gamma & \gamma'\end{pmatrix}$; so by lemma 1.14,

$\rho\begin{pmatrix}\delta & \delta_1 & O & \delta\\ O & -\gamma' & \gamma & O\end{pmatrix} \le \rho(\delta\delta') + \rho(\text{codom }(\gamma\gamma'))$ therefore,

$$\begin{aligned}\rho_\Sigma(\alpha\beta) &\le \rho(\delta\delta') + \rho(\text{codom}(\gamma\gamma')) - \rho(\delta\delta_1) - \rho(\gamma\gamma') + \rho(\text{dom}\alpha)\\ &= \rho(\delta\delta') - \rho(\delta\delta_1) + \rho(\text{codom}\alpha)\end{aligned}$$

since $(\gamma\gamma_1)$ is full with respect to ρ. Since $\text{codom }\alpha = \text{dom }\beta$, $\rho_\Sigma(\alpha\beta) \le \rho_\Sigma(\beta)$.

Again, $\rho\begin{pmatrix}\delta & \delta' & O & -\delta_1\\ O & O & \gamma & \gamma'\end{pmatrix} \le \rho(\gamma\gamma') + \rho(\text{dom }(\delta\delta'))$ and so we find

$$\begin{aligned}\rho_\Sigma(\alpha\beta) &\le \rho(\gamma\gamma') + \rho(\text{dom}(\delta\delta')) - \rho(\delta\delta_1) - \rho(\gamma\gamma_1) + \rho(\text{dom})\\ &= \rho(\gamma\gamma') - \rho(\gamma\gamma_1) + \rho(\text{dom}\alpha) = \rho_\Sigma(\alpha)\end{aligned}$$

since $(\delta\delta_1)$ is full.

For arbitrary α, β, we have the equation:

$$\left(\begin{array}{c|cc|c}\delta & O & O & \delta_1\\ \hline O & \gamma & \gamma_1 & O\end{array}\right)\left(\begin{array}{ccc|c}I & O & O & \beta_1\\ \hline O & I & \alpha_1 & O\\ O & O & \alpha & O\\ \hline O & O & O & \beta\end{array}\right) = \begin{pmatrix}\delta & O & O & \delta'\\ O & \gamma & \gamma' & O\end{pmatrix}$$

from which it is clear that $\rho_\Sigma\begin{pmatrix}\alpha & O\\ O & \beta\end{pmatrix} = \rho_\Sigma(\alpha) + \rho_\Sigma(\beta)$.

Given a f.g. projective P over R_Σ, we wish to define $\rho_\Sigma(P)$ and the natural thing to do is to define $\rho_\Sigma(P) = \rho_\Sigma(e_P)$ where e_P is an idempotent map defined over R_Σ on aniinduced f.g. projective $R_\Sigma \otimes_R Q$ such that the image of e_P is isomorphic to P. We check that this is well-defined.

If e, f are idempotents on $R_\Sigma \otimes_R Q_1$, and $R_\Sigma \otimes_R Q_2$ respectively such that their images are isomorphic to P, there are maps α, β over R_Σ such that $\alpha\beta = e$, and $\beta\alpha = f$; so $\rho_\Sigma(e)$
$\rho_\Sigma(e) = \rho_\Sigma(\alpha\beta) \leq \rho_\Sigma(\beta\alpha\beta\alpha) = \rho_\Sigma(\beta\alpha) = \rho_\Sigma(f) \leq \rho_\Sigma(\alpha\beta\alpha\beta) = \rho_\Sigma(e)$ so that equality must hold and $\rho_\Sigma(P)$ is well-defined.

It is easy to show that the rank ρ_Σ that we have defined on maps between f.g. induced projectives is the inner projective rank function with respect to the projective rank function ρ_Σ that we have just defined on f.g. projectives over R_Σ. Consider the diagram:

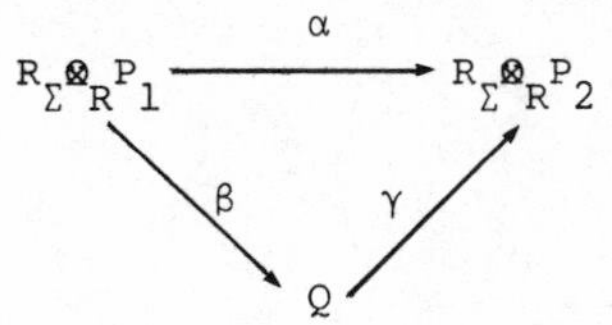

where Q is some f.g. projective. Let $Q \cong (R_\Sigma \otimes_R P)e$; then $\alpha = \bar{\beta} e \bar{\gamma}$ for suitable maps $\bar{\beta}: R_\Sigma \otimes_R P_1 \to R_\Sigma \otimes_R P$, $\bar{\gamma}: R_\Sigma \otimes_R P \to R_\Sigma \otimes_R P_2$. So, $\rho_\Sigma(\alpha) \leq \rho_\Sigma(e) = \rho_\Sigma(Q)$; conversely, we recall that the definition of $\rho_\Sigma(\alpha)$ gives us an equation:

$$9 \qquad (\beta \;\; \beta_1)\begin{pmatrix} I_{\bar{P}} & \alpha_1 \\ 0 & \alpha \end{pmatrix} = (\beta \;\; \beta')$$

where the usual conditions of Cramer's rule hold, and, by definition, $\rho_\Sigma(\alpha) = \rho(\beta \;\; \beta') - \rho(\bar{P})$.

Let $(\beta \;\; \beta') = \gamma\delta$, $\gamma: P' \to P''$, $\delta: P'' \to P'''$ be a minimal factorisation of $(\beta \;\; \beta')$; then

$$10 \qquad \begin{pmatrix} I_{\bar{P}} & 0 \\ 0 & \alpha \end{pmatrix} = (\beta \;\; \beta_1)^{-1}\gamma \quad \delta\begin{pmatrix} I_{\bar{P}} & -\alpha_1 \\ 0 & I \end{pmatrix}$$

consequently, $R_\Sigma \otimes_R P' \cong (R_\Sigma \otimes_R \bar{P}) \oplus Q$, $R_\Sigma \otimes_R P'' \cong (R_\Sigma \otimes_R \bar{P}) \oplus Q''$ and also, $R_\Sigma \otimes_R P''' \cong (R_\Sigma \otimes_R \bar{P}) \oplus Q$; and partitioning (10) according to these decompositions over R, we find the equation:

$$\begin{pmatrix} I_{\bar P} & 0 \\ 0 & \alpha \end{pmatrix} = \begin{pmatrix} I_{\bar P} & \varepsilon_1 \\ \varepsilon_2 & \varepsilon_3 \end{pmatrix} \begin{pmatrix} I_{\bar P} & \phi_1 \\ \phi_2 & \phi_3 \end{pmatrix} = \begin{pmatrix} I_{\bar P} & 0 \\ \mu_1 & \mu_2 \end{pmatrix} \begin{pmatrix} I_{\bar P} & 0 \\ \nu_1 & \nu_2 \end{pmatrix}$$

so, over R_Σ, $\alpha = \mu_2\nu_2$ where $\mu_2 : Q \to Q''$; but we have the equations: $\rho(Q'') = \rho(P'') - \rho(\bar P) = \rho(\beta\ \beta') - \rho(\bar P) = \rho(\alpha)$, which shows that ρ on maps between f.g. induced projectives is the inner projective rank function associated to the projective rank function ρ, and also that there are enough right and left full maps with respect to this rank function for maps between induced projectives.

Let $\alpha : R_\Sigma \otimes_R P_1 \to R_\Sigma \otimes_R P_2$ and $\beta : R_\Sigma \otimes_R P_2 \to R_\Sigma \otimes_R P_3$ be a pair of maps between f.g. induced projectives such that $\alpha\beta = 0$; we have the usual equations deduced from Cramer's rule:

$$(\gamma\ \gamma_1)\begin{pmatrix} I & \alpha_1 \\ 0 & \alpha \end{pmatrix} = (\gamma\ \gamma')\ ,\quad (\delta\ \delta_1)\begin{pmatrix} I & \beta_1 \\ 0 & \beta \end{pmatrix} = (\delta\ \delta')$$

then

$$\left(\begin{array}{ccc|c} \delta & \delta_1 & 0 & 0 \\ 0 & -\gamma' & \gamma & \gamma_1 \end{array}\right) \left(\begin{array}{c|c} & \beta_1 \\ I & \beta \\ & \alpha_1\beta \\ \hline 0 & \alpha\beta \end{array}\right) = \left(\begin{array}{ccc|c} \delta & \delta_1 & 0 & \delta' \\ 0 & -\gamma' & \gamma & 0 \end{array}\right)$$

so $0 = \alpha\beta = (0\ 0\ 0|I)\begin{pmatrix} \delta & \delta_1 & 0 & 0 \\ 0 & -\gamma' & \gamma & \gamma_1 \end{pmatrix}^{-1}\begin{pmatrix} \delta' \\ 0 \end{pmatrix}$ by Malcolmson's criterion

we have an equation over R:

$$\left(\begin{array}{cccccc|c} \delta & \delta_1 & 0 & 0 & 0 & 0 & \delta' \\ 0 & -\gamma' & \gamma & \gamma_1 & 0 & 0 & 0 \\ 0 & 0 & 0 & 0 & \varepsilon_1 & 0 & 0 \\ 0 & 0 & 0 & 0 & 0 & \varepsilon_2 & \phi_2 \\ \hline 0 & 0 & 0 & I & \phi_1 & 0 & 0 \end{array}\right) = \begin{pmatrix} \mu \\ \sigma \end{pmatrix}(\nu|\tau) \qquad (12)$$

where ε_i, μ, ν lie in Σ; so the right nullity of LHS(12) $= \rho(P_3)$. Since ε_1 and ε_2 are full, we deduce from 1.15 that the right nullity of

$\begin{pmatrix} \delta & \delta_1 & 0 & 0 & \delta' \\ 0 & -\gamma' & \gamma & \gamma_1 & 0 \\ 0 & 0 & 0 & I & 0 \end{pmatrix}$ is $\rho(P_3)$ and, in turn, the right nullity of

$\begin{pmatrix} \delta & \delta_1 & 0 & \delta' \\ 0 & -\gamma' & \gamma & 0 \end{pmatrix}$ is $\rho(P_3)$; so, because $\begin{pmatrix} \delta & \delta_1 & 0 \\ 0 & -\gamma' & \gamma \end{pmatrix}$ is right full,

$$\rho\begin{pmatrix} \delta & \delta_1 & 0 & \delta' \\ 0 & -\gamma' & \gamma & 0 \end{pmatrix} = \rho\begin{pmatrix} \delta & \delta_1 & 0 \\ 0 & -\gamma' & \gamma \end{pmatrix} .$$

We see from 1.14 that

$$\rho(\delta\ \delta') + \rho(\gamma\ \gamma') \le \rho\begin{pmatrix} \delta & \delta' & 0 & -\delta_1 \\ 0 & 0 & \gamma & \gamma' \end{pmatrix} = \rho\begin{pmatrix} \delta & \delta_1 & 0 & \delta' \\ 0 & -\gamma' & \gamma & 0 \end{pmatrix} = \rho\begin{pmatrix} \delta & \delta_1 & 0 \\ 0 & -\gamma' & \gamma \end{pmatrix}$$

$$= \rho(\delta\ \delta_1) + (\gamma\ \gamma_1) - \rho(P_1)$$

So $\rho_\Sigma(\alpha) + \rho_\Sigma(\beta) = \rho(\delta\ \delta') - \rho(\delta\ \delta_1) + \rho(P_2) + \rho(\gamma\ \gamma') - \rho(\gamma\ \gamma_1) + \rho(P_1)$
$\le \rho(\delta\ \delta) + \rho(\gamma\ \gamma_1) - \rho(P_1) - \rho(\delta\ \delta_1) + \rho(P_2) - \rho(\gamma\ \gamma_1) + \rho(P_1) = \rho(P_2)$,
which proves the law of nullity for maps between f.g. induced projectives.

Given an arbitrary map $\alpha:Q_1 \to Q_2$ between f.g. projectives over R_Σ, we define $\rho_\Sigma(\alpha)$ to be the inner rank of α with respect to the rank function ρ_Σ on f.g. projectives. If $\alpha:Q_1 \to Q_2$, $\beta:Q_2 \to Q_3$ are maps such that $\alpha\beta = 0$, we find induced f.g. projectives such that $R_\Sigma \otimes_R P_i \cong Q_i \oplus Q_i'$ for $i = 1,2,3$; then $\begin{pmatrix} \alpha & 0 \\ 0 & 0 \end{pmatrix} : Q_1 \oplus Q_1' \to Q_2 \oplus Q_2'$.
$\begin{pmatrix} \beta & 0 & 0 \\ 0 & I_{Q_2'} & 0 \end{pmatrix} : Q_2 \oplus Q_2' \to Q_3 \oplus Q_2' \oplus (Q_3' \oplus Q_2)$ are maps between f.g. induced projectives such that $\begin{pmatrix} \alpha & 0 \\ 0 & 0 \end{pmatrix}\begin{pmatrix} \beta & 0 & 0 \\ 0 & I_{Q_2'} & 0 \end{pmatrix} = 0$ so

$\rho_\Sigma\begin{pmatrix} \alpha & 0 \\ 0 & 0 \end{pmatrix} + \rho_\Sigma\begin{pmatrix} \beta & 0 & 0 \\ 0 & I_{Q_2'} & 0 \end{pmatrix} \le \rho_\Sigma(Q_2) + \rho_\Sigma(Q_2')$ but

$\rho_\Sigma\begin{pmatrix} \alpha & 0 \\ 0 & 0 \end{pmatrix} = \rho_\Sigma(\alpha)$, $\rho_\Sigma\begin{pmatrix} \beta & 0 & 0 \\ 0 & I_{Q_2'} & 0 \end{pmatrix} = \rho_\Sigma(\beta) + \rho_\Sigma(Q_2')$ so we deduce that

$\rho_\Sigma(\alpha) + \rho_\Sigma(\beta) \le \rho_\Sigma(Q_2)$ as we wished to show. So ρ_Σ defines a Sylvester rank function on R_Σ.

Finally, we know that all maps between induced f.g. projectives factor as a right full followed by left full map. So $\begin{pmatrix} \alpha & 0 \\ 0 & 0 \end{pmatrix}$ does (taking the notation of the last paragraph) and it follows that α must since it factors through $\begin{pmatrix} \alpha & 0 \\ 0 & 0 \end{pmatrix}$ and has the same rank as $\begin{pmatrix} \alpha & 0 \\ 0 & 0 \end{pmatrix}$.

It is useful to be able to prove that all f.g. projectives over

R_Σ are stably induced from R in the circumstances of theorem 5.1; this is not in general true, but what does hold is sufficiently good for our purposes.

<u>Theorem 5.2</u> Let R be a ring with a Sylvester projective rank function ρ having enough left and right full maps; let Σ be a collection of full maps with respect to ρ; then any f.g. projective Q over R_Σ satisfies an equation of the form:

$$Q \oplus R_\Sigma^n \oplus Q_0 \cong R_\Sigma \otimes_R P$$

where $\rho(Q_0) = 0$, and P is an f.g. projective over R.

Proof: Let e be an idempotent in $M_m(R_\Sigma)$ such that $R_\Sigma^m e \cong Q$; then by Cramer's rule, we construct the commutative diagram below:

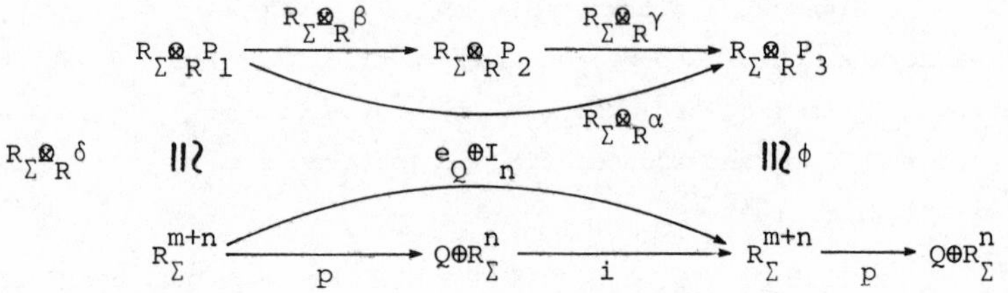

where $\alpha = \beta\gamma$ is a minimal factorisation of α over R, p is the projection of R^{m+n} onto $Q \oplus R^n$ and i is a left inverse to p.

The map $(R_\Sigma \otimes_R \gamma)\phi p : R_\Sigma \otimes_R P_2 \to Q \oplus R^n$ must be surjective; moreover, by the definition of ρ_Σ, $\rho_\Sigma(Q \oplus R^n) = \rho(P_2)$; so in the equation

$$R_\Sigma \otimes_R P_2 \cong Q \oplus R_\Sigma^n \oplus Q_0 \ , \ \rho_\Sigma(Q_0) = 0, \quad \text{as we wished to show.}$$

Incidentally, this shows that the rank function ρ_Σ is the unique extension of the rank function ρ on the image of $K_0(R)$ in $K_0(R_\Sigma)$; for if Q is an f.g. projective such that $\rho_\Sigma(Q) = 0$, there is an equation $Q \oplus R^n \oplus Q_0 \cong R_\Sigma \otimes_R P$, where $\rho(P) = n$, from which it follows that the rank of Q under any rank function extending ρ must be 0; the equation proved in theorem 5.2 for arbitrary Q shows that there is a unique rank function extending ρ.

Let R be a ring with a Sylvester projective rank function ρ

and let Σ_ρ be the collection of full maps with respect to ρ; we call the ring R_{Σ_ρ} the <u>universal localisation of</u> R <u>at</u> ρ, and we shall in general write it as R_ρ. In the case where ρ takes values in $\frac{1}{n}\mathbb{Z}$, we are able to prove a great deal about such a ring.

<u>Theorem 5.3</u> Let R be a ring with a Sylvester projective rank function ρ taking values in $\frac{1}{n}\mathbb{Z}$; then the universal localisation of R at ρ, R_ρ, is a perfect ring with a faithful Sylvester projective rank function ρ and all f.g. projectives over R_ρ are stably induced from R. The kernel of the map $R \to R_\rho$ is the trace ideal of the projectives of rank O, and $R \to R_\rho$ is an honest map.

Proof: We shall need the last half of this theorem in order to prove the first part.

Since ρ takes values in $\frac{1}{n}\mathbb{Z}$ it is clear that there are enough left and right full maps, so we may apply 5.1 and 5.2; in particular R_ρ has a Sylvester projective rank function ρ_Σ taking values in $\frac{1}{n}\mathbb{Z}$, and the map $R \to R_\rho$ must be honest with respect to the pair of rank functions ρ, ρ_Σ.

Let P be an f.g. projective module over R_ρ such that $\rho_\Sigma(P) = 0$; there is an idempotent e in $M_n(R_\rho)$ such that $R^n e \cong P$ and $\rho_\Sigma(e) = \rho_\Sigma(P) = 0$, so $\rho_\Sigma(I_n - e) = n$. By Cramer's rule, we have an equation:

$$(\alpha|\alpha_1)\left(\begin{array}{c|c} I & \beta \\ \hline 0 & I-e \end{array}\right) = (\alpha|\alpha')$$

where $(\alpha\alpha_1)$ is a full map and $(\alpha\alpha')$ is defined over R; since $I_n - e$ has rank n, $(\alpha\alpha')$ is a full map over R, and so, it is invertible over R_Σ; therefore, $I_n - e$ must be invertible and so the identity map. Hence, e = O, P = O, and ρ_Σ is a faithful rank function. All projectives of rank zero over R must be killed by the homomorphism, so by 5.1, the kernel of the map is precisely the trace ideal of the projectives of rank O. By 5.2, all f.g. projectives over R_ρ are stably induced from R. It is easy to see from this that all maps over R_ρ that are full with respect to ρ_Σ are actually invertible. For, if $\alpha: Q_1 \to Q_2$ is a full map with respect to ρ_Σ over R_ρ, there is an integer such that $Q_i \oplus R^n \cong R_\rho \otimes_R P_i$ and $\alpha \oplus I_n : Q_1 \oplus R_\rho^n \to Q_2 \oplus R_\rho^n$ is still a full map with respect to ρ_Σ but this time it is a full map between induced projectives.

By Cramer's rule, we have an equation:

$$(\beta\ \beta_1)\left(\begin{array}{c|c} I & 0 \\ \hline 0 & \begin{matrix}\alpha & 0 \\ 0 & I\end{matrix} \end{array}\right) = (\beta\ \beta')$$

where $(\beta\beta_1)$ is a full map over R and $(\beta\beta')$ is defined over R; since $\begin{pmatrix}\alpha & 0 \\ 0 & I\end{pmatrix}$ is full with respect to ρ_Σ, $(\beta\beta')$ must be full and so it is invertible over R_ρ, which shows that α itself is invertible.

Next, we show that the descending chain condition holds on f.g. left ideals over R_ρ of bounded generating number with respect to ρ_Σ; clearly, this implies that R_ρ is right perfect, and the left perfect condition will follow by symmetry.

If there are any infinite strictly descending chains of f.g. left ideals of bounded generating number, we choose one $I_0 \supset I_1 \supset \ldots.$ where all modules have the same generating number, and this is minimal for such a descending chain to exist. We find f.g. projectives with surjections $P_i \to I_i$ such that $\rho_\Sigma(P_i)$ is the generating number of I_i, and construct the diagram below using projectivity.

$$\begin{array}{ccccccc}
I_0 & \supset & I_1 & \supset & I_2 & \ldots\ldots\ldots \\
\uparrow & & \uparrow & & \uparrow & \\
P_0 & \xleftarrow{\alpha_0} & P_1 & \xleftarrow{\alpha_1} & P_2 & \ldots\ldots\ldots
\end{array}$$

Since $I_i \supsetneq I_{i+1}$, the map we construct from P_{i+1} to P_i cannot be full since it is not even surjective so there is a projective Q_i such that $\rho_\Sigma(Q_i) < \rho_\Sigma(P_i)$ and α_i factors through Q_i; the image of Q_i in I_i is a left ideal I_i' containing I_{i+1} of generating number strictly less than that of I_i, which is contradictory since we may choose a strictly descending chain of left ideals from the sequence $\{I_i'\}$ such that all modules have the same generating number less than that of the sequence $\{I_i\}$.

As noted before, this argument shows by symmetry that R_ρ is also left perfect, so our theorem is complete.

Constructing simple artinian universal localisations

Most of the time, we shall be able to deduce from this last result that R_ρ is actually simple artinian as we shall see in the next two theorems. First of all, however, we give an example of a ring with a faithful Sylvester projective rank function that cannot be embedded in a simple artinian ring at all. We consider the endomorphism ring of the abelian group $C_{p^2} \oplus C_p$ where C_n is the cyclic group of order n. This ring has two non-trivial idempotents corresponding to projection on C_{p^2} and C_p; we assign the ranks $\frac{2}{3}$ to the first, and $\frac{1}{3}$ to the second idempotent. It is a fairly simple matter to check that this gives a Sylvester projective rank function on the ring. It is clear that it cannot be embedded in a simple artinian ring.

The next theorem was proven in a rather different way in (Dicks, Sontag 78); we present it here as a simple corollary of the last theorem.

Theorem 5.4 Let R be a ring with a Sylvester projective rank function ρ taking values in $\mathbb{Z}$; then R_ρ is a skew field and the map $R \to R_\rho$ is honest.

Proof: We know that R_ρ is a perfect ring with a faithful Sylvester projective rank function ρ_Σ taking values in $\mathbb{Z}$; let a,b be elements of R_ρ such that $ab = 0$; then by the law of nullity $\rho_\Sigma(a)$ or else, $\rho_\Sigma(b)$ must be 0, which implies that a or b must be zero. So R_ρ is a domain and a perfect ring which forces it to be a skew field.

This is the classical case, and it is quite sensible to recall at this point some of the examples to which it applies; all our examples are firs, which have the unique rank function on the f.g. projectives since these are free of unique rank. The free algebra on a set X, $k\langle X\rangle$, is a fir, as one sees by writing it as the ring coproduct $\coprod_k k[x_i]$ as x_i runs through X. We generalise this slightly to the ring $E\langle X\rangle$ generated freely by a set X of E-centralising elements; this is a fir by the same argument. There is a further generalisation of this example; let M be an E,E bimodule over the skew field E; we may form the <u>tensor ring</u> on the bimodule M, $E\langle M\rangle$, which is a graded ring whose nth grade has the form $\overset{n}{\otimes} M$. It is shown in (Cohn 71) that this is a fir. The unique universal localisation that is a skew field for these firs are written as $k\langle X\rangle$, $E\langle X\rangle$, and $E\langle M\rangle$.

Our next theorem gives different circumstances under which R_ρ is forced to be a simple artinian ring; this will be exactly what we shall need to show that the simple artinian coproduct exists.

<u>Theorem 5.5</u> Let R be a ring with a unique projective rank function ρ. Assume that ρ is a Sylvester projective rank function whose image is precisely $\frac{1}{n}\mathbb{Z}$; then the universal localisation of R at ρ, R_ρ, is a simple artinian ring of the form $M_n(F)$, where F is a skew field.

Proof: We know that R_ρ is a perfect ring such that all f.g. projectives are stably induced from R; so any rank function on R_ρ is determined by its values on the image of $K_0(R)$. There is a unique rank function on R, so there is a unique rank function on R_ρ. If N is the nil radical of R, we know that R_ρ/N is semisimple artinian; since R_ρ has a unique rank function, R_ρ/N must actually be simple artinian, and since the image of the rank function is precisely $\frac{1}{n}\mathbb{Z}$, and all projectives in R_ρ/N lift to projectives over R_ρ, R_ρ/N must be $M_n(D)$ for some skew field D. The matrix units lift from R_ρ/N to R_ρ, so $R_\rho \cong M_n(R')$, where by Morita equivalence, R' is a perfect ring with a faithful Sylvester projective rank function taking values in $\mathbb{Z}$, so, by the same argument as for 5.4, R' is a skew field, which completes the proof.

This theorem applies well to the ring coproduct of simple artinian rings amalgamating a common simple artinian subring.

<u>Theorem 5.6</u> Let S_1 and S_2 be a couple of simple artinian rings with common simple artinian subring S. Then $S_1 \sqcup_S S_2$ is an hereditary ring with a unique rank function ρ. Therefore, $(S_1 \sqcup_S S_2)_\rho$ is a simple artinian ring, the simple artinian coproduct of S_1 and S_2, amalgamating S. If $S_i \cong M_{n_i}(D_i)$, for skew fields D_i, $(S_1 \sqcup_S S_2)_\rho \cong M_n(D)$, where $n = \text{l.c.m.}\ \{n_1,n_2\}$, and D is a skew field.

Proof: From 3.1, we know that $S_1 \sqcup_S S_2$ has a unique rank function ρ; if $S_i \cong M_{n_i}(D_i)$, the image of ρ is precisely $\frac{1}{n}\mathbb{Z}$ so, by 5.5, $(S_1 \sqcup_S S_2)_\rho \cong M_n(D)$ for some skew field D.

Following Cohn's notation and terminology for the skew field coproduct, we call this ring the simple artinian coproduct of S_1 and S_2 amalgamating S and we denote it by the symbol $S_1 \underset{S}{\circ} S_2$. Of course, we may construct the simple artinian coproduct of finitely many simple artinian rings, S_i, amalgamating a common simple artinian ring in exactly the same way; for infinitely many S_i, we have to be more careful. In this case, the last theorem is proved in the same way provided that l.c.m.$\{n_i\}$ exists, where $S_i \cong M_{n_i}(D_i)$; when this does not exist, our universal localisation at ρ exists but it will be a von Neumann regular ring, that is not simple artinian.

Another generalisation of 5.6 that it is worth mentioning at this point occurs if we amalgamate a semisimple artinian ring rather than a simple artinian ring; by 3.1, we know that if $\alpha:R_0 \to S_i$, $i = 1,2,$ are embeddings of the semisimple artinian ring R_0 in simple artinian rings $S_1 \underset{R_0}{\sqcup} S_2$ has a rank function if and only if α_1 and α_2 induce the same rank function on R_0. When there is a rank function, it is unique and we may apply 5.5 to show that the universal localisation of $S_1 \underset{R_0}{\sqcup} S_2$ at this rank function is a simple artinian ring, which we shall call the simple artinian coproduct of S_1 and S_2 amalgamating R_0, and write as $S_1 \underset{R_0}{\circ} S_2$. It is important to bear in mind that this exists only if the same rank function is induced on R_0 by the maps from R_0 to S_1 and S_2.

Given a ring with a Sylvester projective rank function to $\frac{1}{n}\mathbb{Z}$, we have already seen that it need not arise from a map to a simple artinian ring. However, if our ring is a k-algebra, we can show that there is an honest map to a simple artinian ring inducing the given projective rank function.

<u>Theorem 5.7</u> Let R be a k-algebra with a Sylvester projective rank function ρ to $\frac{1}{n}\mathbb{Z}$; then there is an honest homomorphism from R to a simple artinian ring inducing the rank function ρ.

Proof: We saw in 3.1 that ρ extends to a projective rank function ρ on $M_n(k) \underset{k}{\sqcup} R$, which still takes values in $\frac{1}{n}\mathbb{Z}$; since the rank functions on the factor rings are Sylvester, that on the coproduct must also be Sylvester by 3.7, and the map $R \to M_n(k) \underset{k}{\sqcup} R$ is honest by 3.6. $M_n(k) \underset{k}{\sqcup} R \cong M_n(R')$ and the rank function on the coproduct induces a rank function ρ' on R' which takes values in $\mathbb{Z}$. Consequently, by 5.4, we have a homomorphism from R' to a skew field F that is honest and induces the rank function ρ' on

R'. We have a map $R \to M_n(F)$ which is the composite of honest maps and must be honest.

In particular, we consider when R_ρ is a k-algebra for a ring R with a Sylvester rank function ρ taking values in $\frac{1}{n}\mathbb{Z}$; we see that there is an embedding of R_ρ in $M_n(F)$ for some skew field F, so the nil radical of R must be nilpotent of class at most n.

Intermediate universal localisations

Before leaving these matters for the time being, we shall develop a few results on intermediate localisations of a ring R at some set of maps full with respect to a Sylvester projective rank function ρ. We wish to have useful criteria for the embedding of such rings in the complete localisation at all the full maps. This question is not immediately important, so the reader may wish to skip these results until they are referred to later on.

If R is a ring with a Sylvester projective rank function ρ, and Σ is a collection of full maps with respect to ρ, it is clear that R_ρ must be the universal localisation of R_Σ at ρ_Σ. So R_Σ embeds in R_ρ if and only if ρ_Σ is a faithful projective rank function. There is a useful way of determining when this happens. We define a set of maps Σ to be factor closed, if any full left factor of an element of Σ is invertible in R_Σ.

Theorem 5.8 Let R be a ring with a Sylvester projective rank function ρ taking values in $\frac{1}{n}\mathbf{Z}$; let Σ be a collection of maps full with respect to ρ; then R_Σ embeds in R_ρ if and only if ρ_Σ, the extension of ρ to R_Σ, is a faithful projective rank function, which is true if and only if the lower multiplicative closure of Σ, $\underline{\Sigma}$, is factor closed. The image of R_Σ in R_ρ is always a universal localisation of R.

Proof: We have already seen the first equivalence. Next, suppose that ρ_Σ is a faithful projective rank function and let $\alpha \in \underline{\Sigma}$, $\alpha = \beta\gamma$ where β,γ are full maps $\beta:P \to P'$, $\gamma:P' \to P''$. Over R_Σ, β has a right inverse, and so, $R_\Sigma \otimes_R P' \cong R_\Sigma \otimes_R P \oplus \text{coker}(R_\Sigma \otimes_R \beta)$. $\rho_\Sigma(\text{coker}(R_\Sigma \otimes_R \beta)) = 0$ so it must itself be zero if the rank function is faithful, and, in this case, β has an inverse.

Conversely, suppose that $\underline{\Sigma}$ is factor closed, and let e be an idempotent such that $\rho_{\Sigma}(e) = 0$; we have an equation by Cramer's rule:

$$(\alpha\ \alpha_1) \begin{array}{c|c} I & \beta_1 \\ \hline 0 & e \end{array} = (\alpha\ \alpha')$$

where $(\alpha\alpha_1) \in \underline{\Sigma}$; since $\rho_{\Sigma}(e) = 0$, $\rho(\alpha\alpha') = \rho(\alpha)$, and we must have a minimal factorisation of $(\alpha\alpha')$ of the form $(\alpha\alpha') = \beta(\gamma\gamma)$ where the codomain of β has the same rank as the codomain of α; we have an equation:

$$(\beta\ \alpha_1) \begin{pmatrix} \gamma & 0 \\ 0 & I \end{pmatrix} \begin{pmatrix} I & \beta_1 \\ 0 & e \end{pmatrix} = (\beta\ \alpha_1) \begin{pmatrix} \gamma\ \gamma' \\ \hline 0 \end{pmatrix}$$

where $(\beta\alpha_i)$ is a full left factor of $(\alpha\alpha_1)$ and so, by assumption, is invertible over R_{Σ}; cancelling it over R_{Σ} shows that $e = 0$, which shows that ρ_{Σ} is a faithful projective rank function.

We are left with the last sentence of the theorem; the image of R_{Σ} in R_{ρ} is obtained by killing the f.g. projectives of rank zero, which is the universal localisation of R_{Σ} at idempotent matrices, whose kernels are isomorphic to f.g. projectives of rank 0. By 4.6, this is a universal localisation of R.

It is clear that if $R \to S$ is a ring homomorphism from R to a simple artinian ring inducing a Sylvester projective rank function ρ, then the set of maps Σ between f.g. projectives over R invertible over S is factor closed and saturated; therefore R_{Σ} embeds in R_{ρ} by the last theorem. This remark gives us a useful sufficient condition for a localisation of a ring at a set of maps full with respect to a Sylvester projective rank function to embed in the complete localisation.

<u>Lemma 5.9</u> Let Σ be a collection of full maps with respect to the Sylvester projective rank function ρ over the ring R, and let $R_{\Sigma} \to S$ be an embedding into a simple artinian ring such that the composite map from R to S induces ρ. Then R_{Σ} embeds in R_{ρ}.

Proof: Let $\bar{\Sigma}$ be the collection of maps over R that become invertible over S; then we have the diagram of ring maps:

$$\begin{array}{ccccccc} R & \to & R_{\Sigma} & \to & R_{\overline{\Sigma}} & \to & R_{\rho} \\ & & & \searrow & & \searrow & \\ & & & & & & S \end{array} .$$

The map from $R_{\overline{\Sigma}}$ to S is an embedding, so R_{Σ} embeds in $R_{\overline{\Sigma}}$ which embeds in R_{ρ}.

We can simplify 5.8 in the case that R is a weakly semi-hereditary ring, since in this case, it is easily seen that a set of maps full with respect to a projective rank function are factor closed if and only if their lower multiplicative closure is factor closed.

<u>Theorem 5.10</u> Let R be a weakly semihereditary ring with rank function ρ taking values in $\frac{1}{n}\mathbb{Z}$; let Σ be a collection of maps full with respect to ρ; then R_{Σ} embeds in R_{ρ} if and only if Σ is factor closed.

Proof: It follows from 1.19 that Σ is factor closed if and only if the lower multiplicative closure of Σ is factor closed. 5.8 completes the proof.

Finally, in the case of a faithful rank function on a two sided hereditary ring, we know by 1.23 that we have unique factorisation of full maps into finitely many atomic full maps. Consequently, a universal localisation at a factor closed set of maps is entirely determined by the full maps that are factors of elements of the factor closed set. Further, if Σ_1, Σ_2 are collections of atomic full maps such that any element of one is stably associated to an element of the other, $R_{\Sigma_1} \cong R_{\Sigma_2}$. So, if we wish to classify all the intermediate localisations that embed in the complete one, we can do so by the sets of stable association classes of atomic maps that are exactly those inverted in some universal localisation. We shall show next that any collection is possible.

<u>Theorem 5.11</u> Let R be an hereditary ring with faithful projective rank function ρ taking values in $\frac{1}{n}\mathbb{Z}$; then the intermediate localisations of R that embed in the universal localisation R_{ρ} are in 1 to 1 correspondence with collections of stable association classes of atomic full maps.

Proof: We have shown everything except for the result that if Σ is a

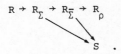

collection of atoms then any atom invertible in R_Σ is stably associated to an element of Σ. In order to deal with this, we look at the maps we defined in the proof of the K-theory exact sequence of chapter 4 for the universal localisation R of R. Let $\underline{T}$ be the category of torsion modules with respect to this complete localisation; we have a well-defined map from isomorphisms between induced f.g. projectives over R_ρ to $K_0(\underline{T})$ and $K_0(\underline{T})$ is the free abelian group on the set of stable association classes of atomic full maps, and our well-defined map sends a full map β to [coker β].

Suppose that we have an atomic full map $\alpha:P \to Q$ such that α^{-1} exists in R_Σ; then, we have an equation of the Cramer rule type, $\alpha(\alpha^{-1}) = I$, which shows that the image of α^{-1} in $K_0(\underline{T})$ is [- coker α]. However, by Cramer's rule for R_Σ, we have an equation:

$$\beta\left(\begin{array}{c|c} I & \alpha' \\ \hline 0 & \alpha^{-1} \end{array}\right) = \beta'$$

where β lies in the lower multiplicative closure of Σ, and β' is a full map. So, the image of α^{-1} is [coker β'] - [coker β]. Therefore, since our map is well-defined, [coker α] + [coker β'] = [coker β]. Since each side is a positive sum of generators of a free abelian group, and [coker α] is a generator, it equals one of the generators involved in the right hand side, which all have the form [coker α_i] for α_i in Σ; so, α is stably associated to an element of Σ.

In the course of the proof of the theorems in this chapter, we have often tried to prove that all the f.g. projectives over a universal localisation are stably induced; it is often useful to be able to show that all f.g. projectives are actually induced; we end this chapter with a consideration of this problem.

Let R be a ring with a Sylvester projective rank function ρ and let Σ be a collection of maps between f.g. projectives full with respect to ρ; we say that Σ is <u>factor complete</u> if it is factor closed and if α and β are maps defined over R such that $\alpha\beta$ lies in Σ, there exists α', a map between induced f.g. projectives over R_Σ, such that $\begin{pmatrix}\alpha \\ \alpha'\end{pmatrix}$ is invertible over R_Σ.

<u>Theorem 5.12</u> Let R be a ring with a Sylvester projective rank function ρ;

let Σ be a lower multiplicatively closed factor complete set of maps between f.g. projectives; then all f.g. projectives over R_Σ are induced from R.

Proof: Let e be an idempotent in $M_n(R_\Sigma)$; by Cramer's rule, there exists an equation:

$$1 \qquad (\alpha\ \alpha_1)\begin{pmatrix} I_P & \beta \\ 0 & e \end{pmatrix} = (\alpha\ \alpha')$$

where $(\alpha\ \alpha_1) \in \Sigma$, $(\alpha\ \alpha'):P_0 \to P \oplus R^n$ and $(\alpha\ \alpha')$ is defined over R, $(\alpha\ \alpha'):P_0 \to P \oplus R^n$.

Let $(\alpha\ \alpha') = \gamma(\delta\ \delta')$ be a minimal factorisation over R; $\gamma:P_0 \to Q$ and $(\delta\ \delta'):Q \to P \oplus R^n$; then $(\gamma\ \alpha_1)\begin{pmatrix} \delta & 0 \\ 0 & I \end{pmatrix} = (\alpha\ \alpha_1)$ so, by factor completeness, there exists $(\varepsilon_1\ \varepsilon_2):R_\Sigma \otimes_R P' \to R_\Sigma \otimes_R Q \oplus R^n_\Sigma$ such that $\begin{pmatrix} \gamma & \alpha_1 \\ \varepsilon_1 & \varepsilon_2 \end{pmatrix}$ is invertible.

Counting ranks shows that $\rho(P') = \rho(Q) + n - \rho(P_0)$ and also that $\rho(Q) = \rho(P) + \rho_\Sigma(e)$; therefore, $\rho(P') = \rho_\Sigma(e) + \rho(P) + n - \rho(P_0) = \rho_\Sigma(e)$.

We intend to show that there is a map from $R_\Sigma \otimes_R P'$ onto R^n_e, and hence that they must be isomorphic since there are no projectives of rank 0 by the factor closure of Σ.

First we rewrite 1 as:

$$(\gamma\ \alpha_1)\begin{pmatrix} \delta & 0 \\ 0 & I \end{pmatrix}\begin{pmatrix} I_\rho & \beta \\ 0 & e \end{pmatrix} = (\gamma\ \alpha_1)\begin{pmatrix} \delta & \delta' \\ 0 & 0 \end{pmatrix}$$

so

$$(\gamma\ \alpha_1)\begin{pmatrix} 0 & \delta\beta \\ 0 & e \end{pmatrix} = 0 \qquad (\gamma\ \alpha_1)\begin{pmatrix} 0 & \delta\beta e \\ 0 & e \end{pmatrix} = 0$$

$$\Longrightarrow (I_{P_0}\ 0)\begin{pmatrix} \gamma & \alpha_1 \\ \varepsilon_1 & \varepsilon_2 \end{pmatrix}\begin{pmatrix} I & \delta\beta \\ 0 & I \end{pmatrix}\begin{pmatrix} 0 & 0 \\ 0 & e \end{pmatrix} = e$$

On the one hand, $\begin{pmatrix}\gamma & \alpha_1\\ \varepsilon_1 & \varepsilon_2\end{pmatrix}\begin{pmatrix}I & \delta\beta\\ O & I\end{pmatrix}\begin{pmatrix}O & O\\ O & e\end{pmatrix}$ has image $R^n e$ since $\begin{pmatrix}\gamma & \alpha_1\\ \varepsilon_1 & \varepsilon_2\end{pmatrix}\begin{pmatrix}I & \delta\beta\\ O & I\end{pmatrix}$ is invertible; on the other hand, it is a map from $R_\Sigma \otimes_R (P_O \oplus P')$ vanishing on $R_\Sigma \otimes_R P_O$; so it defines a surjective map from $R_\Sigma \otimes_R P'$ to R^n_e, which completes our proof.

We may refine this a little in the case where R is weakly semihereditary.

Theorem 5.13 Let R be weakly semihereditary with a projective rank function ρ; let Σ be a factor complete collection of maps between f.g. projectives over R full with respect to ρ; then the lower multiplicative closure of Σ, $\underline{\Sigma}$, is also factor complete and so, all f.g. projectives over R_Σ are induced from R.

Proof: It is sufficient to show that if Σ is factor complete then $\Sigma_2 = \{\begin{pmatrix}\alpha_1 & O\\ \beta & \alpha_2\end{pmatrix} : \alpha_i \in \Sigma\}$ is also factor complete; the rest follows by induction. So, suppose that $\begin{pmatrix}\gamma_1\\ \gamma_2\end{pmatrix}(\delta_1\ \ \delta_2) = \begin{pmatrix}\alpha_1 & O\\ \beta & \alpha_2\end{pmatrix}$ then $\gamma_1\delta_2 = O$, so by the weak semihereditary property, there exists a decomposition of $\mathrm{codom}(\gamma_1) = \mathrm{dom}(\delta_2)$ so that the product $\gamma_1\delta_2$ is trivially O; we rewrite the above equation with respect to this partition of $\mathrm{codom}(\gamma_1)$:

$$\begin{pmatrix}\gamma_{11} & O\\ \gamma_{21} & \gamma_{22}\end{pmatrix}\begin{pmatrix}\delta_{11} & O\\ \delta_{21} & \delta_{22}\end{pmatrix} = \begin{pmatrix}\alpha_1 & O\\ \beta & \alpha_2\end{pmatrix}$$

By factor completeness, there exists γ' and γ'' such that $\begin{pmatrix}\gamma_{11}\\ \gamma'\end{pmatrix}$ and $\begin{pmatrix}\gamma_{22}\\ \gamma''\end{pmatrix}$ are invertible: then $\begin{pmatrix}\gamma_{11} & O\\ \gamma_{21} & \gamma_{22}\\ \gamma' & O\\ O & \gamma''\end{pmatrix}$ is invertible.

Finally, there is the following special case that we shall need later on.

Theorem 5.14 Let R be a hereditary ring with a faithful projective rank function ρ; let $\alpha: P_O \to P_1$ be an atomic full map with respect to ρ such that if $\operatorname{im}\alpha \subset Q \subset P_1$ then $\operatorname{im}\alpha$ is a direct summand of Q; then all f.g. projectives over R_α are induced from R.

Proof: It is enough to show that $\{\alpha\}$ is factor complete; it is certainly factor closed, and the remaining part of the definition of factor completeness is trivial.

6. HOMOMORPHISMS FROM HEREDITARY TO VON NEUMANN REGULAR RINGS

In the last chapter, we investigated the universal localisation of a ring at a Sylvester rank function taking values in $\frac{1}{n}\mathbb{Z}$; the resulting ring is always a perfect ring. Since the degree of nilpotence of the radical may grow with the integer n, it is likely that if the rank function takes values in the real numbers, we are not going to be able to say a great deal about the universal localisation at the rank function in general. This suggests that rather than investigating epimorphisms we should investigate homomorphisms, and we shall see that every rank function on an hereditary ring arises from a homomorphism to a von Neumann regular ring with a rank function. Under suitable hypotheses we shall be able to show that the universal localisation at the rank function is von Neumann regular.

In chapter 1, we showed that over a two-sided χ_o-hereditary ring with rank function ρ, every map factors as a right full followed by a left full map; we described this as having enough right and left full maps. We shall require a complementary condition; we say that R has enough full maps with respect to a Sylvester rank function ρ if every left full map is a left factor of a full map and every right full map is a right factor of a full map.

Theorem 6.1 Let R be a ring with a Sylvester rank function ρ such that R has enough left full, right full and full maps with respect to ρ. Then the universal localisation of R at ρ, R_ρ, is a von Neumann regular ring. All f.g. projectives over R are stably induced from R and the rank function extends to R_ρ. The kernel of the homomorphism from R to R_ρ is the trace ideal of the f.g. projectives of rank zero.

Proof: We have seen all of this result in 5.1 and 5.2, except for the fact that R_ρ is a von Neumann regular ring. To show this, we need to show that every principal left ideal of R_ρ is a direct summand of R_ρ. Let $a \in R_\rho$, then right multiplication by a is stably associated to some map $R_\rho \otimes_R \alpha$, where $\alpha : P \to Q$ is a map defined over R. $\alpha = \beta\gamma$, where β is right full and γ is left full. Since β is right full, it is a right factor of some full map which has an inverse over R_ρ, so $R_\rho \otimes_R \beta$ has a left inverse, that is, it is a split surjection. Similarly, $R_\rho \otimes_R \gamma$ has a right inverse, and must be a split injection. The consequence of this is that the image of $R_\rho \otimes_R \alpha$ is a direct summand of $R_\rho \otimes_R Q$, and so, the cokernel of $R_\rho \otimes_R \alpha$ is f.g. projective. This is the cokernel of right multiplication by a, however, so $R_\rho a$ is a direct summand of R_ρ as we wished to show.

Of course, it may be hard to check whether there are enough full maps; however, we have already seen that over an χ_o-hereditary ring there are enough right and left full maps. We should like to be able to embed any k-algebra with a Sylvester rank function ρ honestly in another k-algebra with a Sylvester rank function such that the second k-algebra has enough full maps. It turns out that we can do this by adjoining a large number of generic maps.

<u>Theorem 6.2</u> Let R be a k-algebra with a Sylvester rank function ρ taking values in the reals; then the embedding of R in $R \sqcup_k k\langle X\rangle$, where X is an infinite set is an honest map for the rank functions ρ on R, and (ρ, r) on the coproduct $R \sqcup_k k\langle X\rangle$, where r is the unique rank function on $k\langle X\rangle$. Further, $R \sqcup_k k\langle X\rangle$ has enough full maps with respect to (ρ, r).

Proof: Since all projectives are induced from R, we take liberties with the notation by writing P for $(R \sqcup_k k\langle X\rangle) \otimes_R P$, and ρ for (ρ, r). The idea of the proof is a fairly simple one; let $\alpha : P \to Q$ be a left full map with respect to ρ, defined over $R \sqcup_k k\langle X\rangle$. Then it is actually defined over $R \sqcup_k k\langle Y\rangle$, where Y is a finite subset of X. We may use the elements of $X-Y$ to define a generic map from Q to P over $R \sqcup_k k\langle X\rangle$ and the composition of this with α should be a full map with respect to ρ, since any other answer would involve some kind of degeneracy. A dual argument deals with right full maps.

There is some integer n such that both P and Q are n-generator projectives, and neither is of rank n with respect to ρ; so there are idempotents e_P, e_Q in $M_n(R)$ such that $R^n e_P \cong P$, and $R^n e_Q \cong Q$ so that our left full map $\alpha : P \to Q$ may be represented by a matrix

$a \in M_n(R \sqcup_k k\langle Y\rangle)$ such that $e_P a = a = a e_Q$. For convenience in writing the proof, we pass by Morita equivalence to the ring $M_n(R \sqcup_k k\langle Y\rangle) \sqcup_k k[x] \cong M_n(R \sqcup_k k\langle Y\rangle \sqcup_k k\langle x_{ij}\rangle)$, where $i,j = 1 \to n$, and the map sends x to the matrix (m_{ij}). We call the ring $M_n(R \sqcup_k k\langle Y\rangle) R_1$, and $R_1 \sqcup_k k[x]$ is called S. This may be regarded as a subring of $M_n(R \sqcup_k k\langle X\rangle)$ by identifying the set $\{x_{ij}: i,j = 1 \to n\}$ with some subset of $X - Y$. The rank function ρ induces by Morita equivalence, a rank function ρ_n on $M_n(R \sqcup_k k\langle X\rangle)$ which in turn induces rank functions that we shall still call ρ_n on R_1 and S, and all maps mentioned above are honest since all maps represent one ring as a factor in a coproduct over k of the other. $\rho_n(R_1 e_P)$, $\rho_n(R_1 e_Q) < 1$ since, by Morita equivalence, they must be respectively $\rho(P)/n$ and $\rho(Q)/n$. The element a represents the left full map $\alpha: P \to Q$ so that right multiplication by a is a left full map with respect to ρ_n from $R_1 e_P$ to $R_1 e_Q$, and since the embedding of $R_1 e_Q$ in R_1 is split injective and so left full, right multiplication by a defines a left full map from $R_1 e_P$ to R_1; so for all left ideals of R_1 containing a, $\rho_n(I) \geq \rho_n(R_1 e_P)$.

Over the ring $S = R_1 \sqcup_k k[x]$ the map xe_P is a kind of generic map from R_1 to $R_1 e_P$ so we consider the map axe_P from Se_P to itself, which we intend to show is a full map over S; once we have this, we know that it defines a full map over $M_n(R \sqcup_k k\langle X\rangle)$ by our remark that the inclusion is honest, and so the map Morita equivalent to it over $R \sqcup_k k\langle X\rangle$ is a full map having $\alpha: P \to Q$ as a left factor. In order to deal with this, we shall need the details of the coproduct theorems.

What we need to show is that any f.g. left ideal inside Se_P containing axe_P has generating number at least $\rho_n(Se_P)$ with respect to ρ_n. Let M be such a submodule of Se_P; then, if $g\rho_n(M) \leq \rho_n(Se_P) < 1$, we find that in the decomposition given by 2.7, $M \cong S \otimes_k M_0 \oplus S \otimes_{R_1} M_1 \oplus S \otimes_{k[x]} M_2$, where we take R_0 of 2.7 to be k, R_1 to be R_1 and R_2 to be $k[x]$, M_0 and M_2 must be 0, for if not the generating number of M would be greater than 1 by 3.3. Therefore, $M \cong S \otimes_{R_1} M_1$.

We choose the basis $1, x, x^2, \ldots$ for $k[x]$ and order it by $1 < x < x^2 < \ldots$; we well-order some basis for R_1 over k containing the element e_P as smallest element. Then by theorem 2.7, M_1 is the R_1-submodule of M consisting of the elements whose 1-support does not contain the 1-leading term of some non-1-pure element of M; if $axe_P \in M_1$, its

1-support xe_P must be the 1-leading term of some non-1-pure element of M, and the nature of the ordering of $k[x]$ forces the form of this element to be $xe_P + r_1$ where r_1 lies in R_1e_P. However, M_2 is empty, so $xe_P + r_1$ does not lie in M_2 and we see that e_P must lie in M. For, some element in the 2-support of $xe_P + r_1$ must be the 2-leading term of some non-2-pure element of M. If this 2-leading term is e_P, the element must actually be e_P since we chose e_P to be the smallest element of a basis of R_1. In the contrary case we reduce the support of r_1 obtaining $xe_P + r_2$. The same argument applies but we cannot produce an infinite sequence of terms $xe_P + r_n$ where the maximal element in the support of r_n continues to decrease since our ordering is a well-ordering, so eventually e_P is forced to be in M, and therefore $g\rho_n(M) = \rho_n(Se_P)$.

This leaves the case that axe_P lies in M_1; we recall from theorem 2.6 that the structure of Se_P as an R_1-module has the form $R_1e_P \oplus R_1xe_P \oplus B$, where B is a basic module. We project M_1 onto the direct summand R_1xe_P according to this decomposition. R_1xe_P is a free module on the generator xe_P; the image of M_1 contains axe_P; we have already noted that any left R_1 ideal containing a has generating number at least $\rho_n(R_1e_P) = \rho_n(Se_P)$, since a defines a left full map from R_1e_P to R_1; so the generating number of M_1 is at least $\rho_n(R_1e_P)$ and so by 3.3 the generating number of M is at least $\rho_n(R_1e_P)$, which is just $\rho_n(Se_P)$, so that axe_P must be a full map, as we wished to show.

This shows that every left full map over $R \sqcup_k k\langle X\rangle$ is a left factor of a full map; the dual result must hold for right full maps by a dual argument.

An immediate corollary of 6.1 and 6.2 is the following theorem.

<u>Theorem 6.3</u> Let R be an χ_o-hereditary k-algebra with a rank function ρ taking values in the real numbers; then there is an honest map from R to a von Neumann regular ring V with a rank function ρ_V.

Proof: First of all, we form the ring coproduct $R \sqcup_k k\langle X\rangle$ which is χ_o-hereditary by the remarks after 2.10, so, by 1.16, it has enough right and left full maps; by 6.2, it has enough full maps. Therefore, by 6.1, the universal localisation of $R \sqcup_k k\langle X\rangle$ at the rank function ρ is a von Neumann regular ring V. The map $R \to R \sqcup_k k\langle X\rangle$ and the map $R \sqcup_k k\langle X\rangle \to V$ are both honest, so their composite is also honest.

We can say a little about the monoid of f.g. projectives of V, when it is constructed in the manner used in 6.3.

Theorem 6.4 Let R be a ring with a Sylvester rank function ρ whose image is the subgroup A of the additive reals. Let V be the universal localisation of $R \sqcup_k k\langle X\rangle$ at the rank function ρ extended to $R \sqcup_k k\langle X\rangle$; then $P_{\oplus}(V)$ is naturally identified with the positive cone of A.

Proof: Certainly, the image of ρ on V is the image of ρ on $R \sqcup_k k\langle X\rangle$ by theorem 5.1, which is the image of ρ on R. The rank function induces a map from $P_{\oplus}(V)$ to the positive cone of A, which we wish to show is an isomorphism.

Suppose that P and Q are f.g. projectives over V such that $\rho_V(P) \leq \rho_V(Q)$; then there are idempotents e_P and e_Q in $M_n(V)$ where n is a suitable integer larger than $\rho_V(P)$ such that $V^n e_P \cong P$, and $V^n e_Q \cong Q$. e_P and e_Q must lie in $(R \sqcup_k k\langle Y\rangle)_\rho$ for some finite subset Y of X.

We look at the ring $S = M_n(R \sqcup_k k\langle Y\rangle) \sqcup_k k[x] \cong M_n((R \sqcup_k k\langle Y\rangle) \sqcup_k k\langle x_{ij}\rangle)$ as in the last theorem, which may be defined to be a subring of $M_n(V)$ such that the inclusion of it in $M_n(V)$ is honest by taking $\{x_{ij}\}$ to be a subset of $X - Y$. The map from Se_P to Se_Q given by right multiplication by $e_P x e_Q$ is seen to be a full map by the argument of 6.2, if $\rho_V(P) = \rho_V(Q)$; it is left full if $\rho_V(P) < \rho_V(Q)$. Full maps become isomorphisms over V, and left full maps become split injective, so we see that if P and Q are f.g. projectives of the same rank over V, they are isomorphic, whilst if P has rank less than that of Q it is a direct summand of Q. The first statement shows that the map from $P_{\oplus}(V)$ to the positive cone of A is injective; since all elements of A are differences of elements of the image of $P_{\oplus}(V)$, the second statement shows that the map is surjective.

7. HOMOMORPHISMS FROM RINGS TO SIMPLE ARTINIAN RINGS

Introduction

Now that we have studied special homomorphisms from hereditary rings to simple artinian rings, we are in a good position to study all possible homomorphisms from an arbitrary ring to simple artinian rings. In order to see what type of theory to look for, we should look at the special case of homomorphisms from an arbitrary ring to skew fields, which were classified by Cohn in chapter 7 of (Cohn 71).

First of all, given a homomorphism $\phi:R \to F$ from a ring R to a skew field F, we can talk of the skew subfield of F generated by R; then we regard two homomorphisms $\phi_i:R \to F_i$, $i = 1,2$, as equivalent if the skew subfields of F_i are isomorphic as R-rings. We should like to be able to characterise the equivalence classes in some way; this we do by using the notion of the <u>singular kernel</u> of the homomorphism $\phi:R \to F$; this is the set of square matrices over R that are singular over F. Such a set of matrices P must satisfy the following axioms;

1/ it includes all non-full matrices; these are the n by n matrices for arbitrary n, that can be written as the product of an n by $(n-1)$ and an $(n-1)$ by n matrix;

2/ $1 \notin P$;

3/ $\begin{pmatrix} A & 0 \\ 0 & B \end{pmatrix} \in P$ if and only if A or $B \in P$;

4/ if $A = (a_{ij})$ and $B = (b_{ij})$, $A,B \in P$, and $a_{ij} = b_{ij}$ $i \neq k$, then $(c_{ij}) \in P$ where $c_{ij} = a_{ij}$ for $i \neq k$, and $c_{kj} = a_{kj} + b_{kj}$; similarly, if $a_{ij} = b_{ij}$ $j \neq k$, then $(d_{ij}) \in P$, where $d_{ij} = a_{ij}$ for $j \neq k$ and $c_{ik} = a_{ik} + b_{ik}$.

Any set of matrices over a ring satisfying axioms 1 to 4 is called a <u>prime matrix ideal</u>; Cohn showed that the equivalence classes of homomorphisms from a ring to skew fields are in 1 to 1 correspondence to prime matrix ideals, where an equivalence class is paired with the associated singular kernel. We shall present a proof of these results later in this chapter.

There are two problems involved in a generalisation of this theory: first of all, we cannot talk of the simple artinian subring of a simple artinian ring generated by a subring since there need not be a unique minimal simple artinian ring containing a given subring; therefore, we shall have to find a new way to define an equivalence relation on homomorphisms from a ring to simple artinian rings; secondly, we must find some analogue of the prime matrix ideals that applies to simple artinian rings and not just to skew fields.

The first problem can be met in a fairly simple way; Bergman proposed that two homomorphisms $\phi_i : R \to S_i$, $i = 1,2$, where S_i is simple artinian, should be regarded as equivalent if there is a commutative diagram:

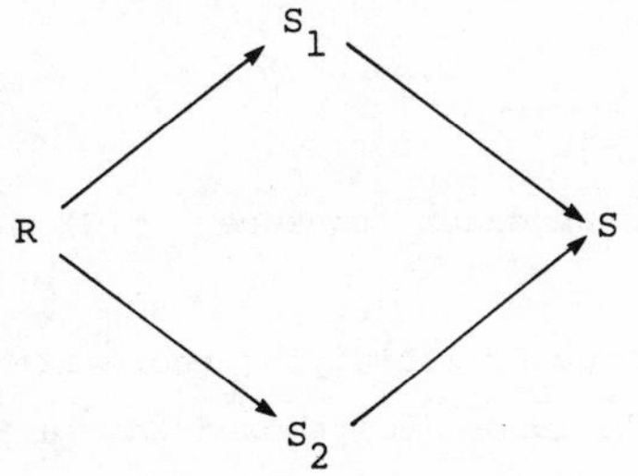

where S is simple artinian. If there is such a commutative diagram, we write $\phi_1 \sim \phi_2$. This relation reduces to the standard one if S_i is a skew field. This is easily shown to be an equivalence relation using the simple artinian coproduct.

<u>Lemma 7.1</u> $\sim$ is an equivalence relation.

Proof: Certainly, it is reflexive and symmetric by definition. If we have homomorphisms $\phi_i : R \to S_i$, $i = 1$ to 3, and $\phi_1 \sim \phi_2$, $\phi_2 \sim \phi_3$, we construct a commutative diagram of ring homomorphisms:

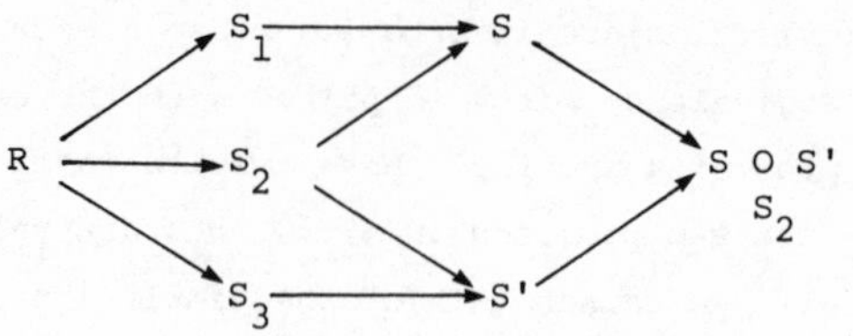

from which it is clear that $\sim$ is also a transitive relation.

Next, we have to see what concept should replace the prime matrix ideal. Here, we use an idea due, in the case of homomorphisms from a ring to skew fields, to Malcolmson (80). If we have a homomorphism $\phi: R \to S \cong M_n(E)$, where E is a skew field, we can define a rank on f.p. modules over R by the formula $\rho(M) = \rho_S(M \otimes_R S)$, which takes values in $\frac{1}{n}\mathbb{Z}$. Such a rank function must satisfy the following axioms, which simply express that $\otimes_R S$ is a right exact functor:

1/ $\rho(R^1) = 1$;

2/ $\rho(A \oplus B) = \rho(A) + \rho(B)$;

3/ if $A \to B \to C \to 0$ is an exact sequence, $\rho(C) \le \rho(B) \le \rho(A) + \rho(C)$.

A rank function on f.p. modules satisfying these axioms is called a Sylvester module rank function. Malcolmson (80) showed that a Sylvester module rank function taking values in $\mathbb{Z}$ is equivalent information on the ring to a prime matrix ideal. The main theorem of this chapter is that the equivalence classes of homomorphisms from a k-algebra to simple artinian rings are in 1 to 1 correspondence Sylvester module rank functions taking values in $\frac{1}{n}\mathbb{Z}$ for some n.

In order to prove this theorem and also to show how the notion of a prime matrix ideal links with the concept of a Sylvester module rank function, we need another type of rank. A homomorphism from a ring R to a simple artinian ring S also determines a rank on maps between f.g. projectives; if $\alpha: P \to Q$ is a map between f.g. projectives over R, we define $\rho(\alpha) = \rho_S(\alpha \otimes_R S)$ where $\rho_S(\beta)$ for a map β between f.g. modules over S is the rank as S-module of the image. A rank on maps induced by a homomorphism to a simple artinian ring must satisfy the following axioms:

1/ $\rho(I_1) = 1$;

2/ $\rho\begin{pmatrix}\alpha & 0\\ 0 & \beta\end{pmatrix} = \rho(\alpha) + \rho(\beta)$;

3/ $\rho\begin{pmatrix}\alpha & 0\\ \alpha & \beta\end{pmatrix} \geq \rho(\alpha) + \rho(\beta)$;

4/ $\rho(\alpha\beta) \leq \rho(\alpha), \rho(\beta)$.

A rank function on maps satisfying these axioms is called a Sylvester map rank function.

The notion of a Sylvester map rank function is equivalent to that of a Sylvester module rank function in the following way. First, we suppose that we have a Sylvester map rank function, ρ; we extend this to a Sylvester module rank function by $\rho(\text{coker } \alpha) = \rho(I_Q) - \rho(\alpha)$ for a map $\alpha: P \to Q$; next, we suppose that we have a Sylvester module rank function, ρ, from which we define a Sylvester map rank function by $\rho(\alpha) = \rho(Q) - \rho(\text{coker } \alpha)$ for a map $\alpha: P \to Q$. It is not hard to check that these functions are well-defined and satisfy the axioms required. As we have seen, Sylvester module and map rank functions are equivalent notions; we shall therefore usually refer to a Sylvester rank function, ρ, which is defined on f.p. modules and on maps between f.g. projectives and restricts respectively to a Sylvester module and a Sylvester map rank function.

In certain situations, we shall be able to construct a rank function on matrices taking values in $\frac{1}{n}\mathbb{Z}$ for some n that satisfies axioms 1 to 4 for a Sylvester map rank function; we shall call such a function a Sylvester matrix rank function. There is the following useful observation.

Lemma 7.2 A Sylvester matrix rank function extends uniquely to a Sylvester map rank function, having values in the same subset of the additive group of $\mathbb{R}$.

Proof: Given a Sylvester matrix rank function, ρ, we define a Sylvester module rank function by $\rho(\text{coker } A) = n - \rho(A)$, where $A: {}^mR \to {}^nR$. That this defines a Sylvester module rank function is the same proof as was needed to show that a Sylvester map rank function determines a Sylvester module rank function. In turn, the Sylvester module rank function determines a Sylvester map rank function that extends the Sylvester matrix rank function.

This allows us to see the equivalence between prime matrix ideals and Sylvester rank functions taking values in $\mathbb{Z}$ in the following way. We construct from the prime matrix ideal, P, a Sylvester matrix rank function by defining $\rho(A) = n$, where n is the maximal integer such that A contains an n by n minor not in P; this determines a Sylvester rank function as we discussed before. Conversely, if we have a Sylvester rank function, ρ, which takes values in $\mathbb{Z}$ we may define a prime matrix ideal by $P = \{A: \rho(A) < n,$ where A is an n by n matrix$\}$. We shall leave the checking of the details to the reader.

Characterising the homomorphism by the rank function

We take our first step by describing our previous equivalence relation on homomorphisms to simple artinian rings by the associated Sylvester map rank functions. In order to do this, we need to find out more exactly what a Sylvester map rank function induced by a homomorphism from R to a simple artinian ring S tells us about the functor $\otimes_R S$.

We have already seen that a Sylvester map rank function determines a Sylvester module rank function, and it is clear that this is the Sylvester module rank function induced by the homomorphism from R to S, for if M is an f.p. module over R with presentation $P \xrightarrow{\alpha} Q \to M \to 0$, then denoting the Sylvester map and module rank functions induced by the homomorphism from R to S by ρ, we have the equation $\rho(\alpha) + \rho(M) = \rho(Q) = \rho(I_Q)$. We can further determine the rank of $M \otimes_R S$ for an arbitrary f.g. module M over R in terms of the Sylvester module rank function. For every f.g. module M can be represented as the direct limit of a family of f.p. modules $\{N_i\}$ where all the maps are surjective, and the rank of $M \otimes_R S$ is equal to the minimal rank of some $N_i \otimes_R S$, that is, the minimal value of $\rho(N_i)$.

Given a map between f.g. modules $\alpha: M \to N$, we may determine the rank of $\alpha \otimes_R S$ by the formula $\rho(\alpha \otimes_R S) = \rho(\text{coker } \alpha \otimes_R S) = \rho(N \otimes_R S)$. Finally, if we have a map $\alpha: M \to N$ where only M is f.g., we can determine the rank of $\alpha \otimes_R S$ in the following way; we write N as the directed union of all the f.g. submodules of N that contain the image of α, $N = \bigcup_j N_j$; then the rank of $\alpha \otimes_R S$ is the minimal rank of the maps from M to N_j for varying j induced by α from M to N.

We see that if two homomorphisms $\phi_1: R \to S_1$ and $\phi_2: R \to S_2$ induce the same Sylvester map rank function, then they must agree numerically in all the ways determined above; as we shall see this happens because there is a commutative diagram of ring homomorphisms:

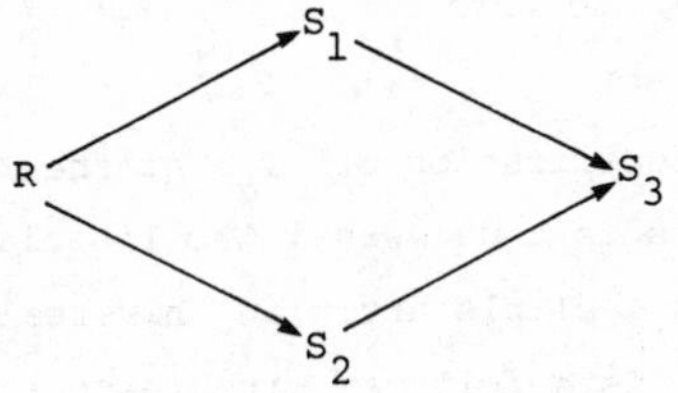

where S_3 is simple artinian.

<u>Theorem 7.3</u> Let R be a ring with two homomorphisms $\phi_1 : R \to S_1$ and $\phi_2 : R \to S_2$ from R to artinian rings S_1 and S_2 that induce the same Sylvester map rank functions; then there is a commutative diagram:

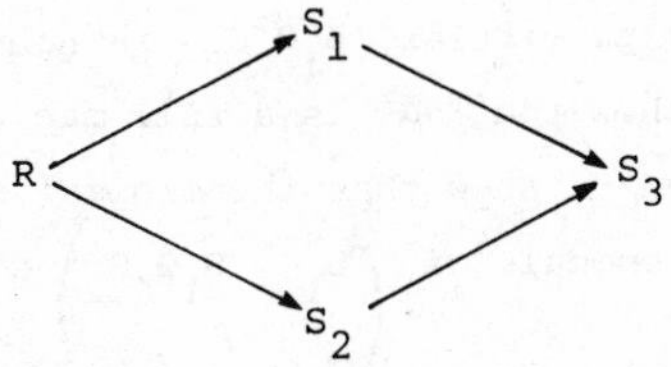

where S_3 is simple artinian. Conversely, if there is such a commutative diagram, S_1 and S_2 induce the same Sylvester map rank function on R.

Proof: The last sentence is clear, since they must both induce the same Sylvester map rank function as S_3.

We wish to find a non-zero homomorphism from the ring $S_1 \underset{R}{\sqcup} S_2$ to a simple artinian ring. At first sight, this appears a hopeless task until one notices that this is an hereditary ring, as will become clear in a moment. After that, it is simply a matter of showing that the ring has unbounded generating number.

In order to study the ring, we consider the upper triangular matrix ring $T = \begin{pmatrix} S_1 & S_1 \otimes_R S_2 \\ 0 & S_2 \end{pmatrix}$.

If we adjoin the universal inverse to the map α from $\begin{pmatrix} 0 & 0 \\ 0 & S_2 \end{pmatrix}$ to $\begin{pmatrix} S_1 & S_1 \otimes_R S_2 \\ 0 & 0 \end{pmatrix}$ defined by left multiplication by $\begin{pmatrix} 0 & 1 \otimes_R 1 \\ 0 & 0 \end{pmatrix}$ the ring we obtain is isomorphic to $M_2(S_1 \underset{R}{\sqcup} S_2)$, as we saw in 4.10. Consequently, $S_1 \underset{R}{\sqcup} S_2$ is hereditary, and by 5.1, it has a rank function (which must be unique) when this map is full with respect to the rank function on T, ρ, that assigns the rank $\frac{1}{2}$ to both $\begin{pmatrix} 0 & 0 \\ 0 & S_1 \end{pmatrix}$ and $\begin{pmatrix} S_1 & S_1 \otimes_R S_2 \\ 0 & 0 \end{pmatrix}$. If this happens, the universal localisation of T_α at the rank function is simple artinian by 5.5; but this is a universal localisation of $M_2(S_1 \sqcup_R S_2)$; therefore, $S_1 \sqcup_R S_2$ has a simple artinian universal localisation. So, it remains to show that α is a full map with respect to ρ.

First of all, we show $S_1 \otimes_R S_2$ is not zero. The rank of the map $\phi_1 : R \to S_1$ as a map of R-modules is 1, where by rank, we mean the rank function induced by the homomorphisms to S_1 and S_2; we show this by considering the map obtained from it by tensoring over R with S_1; we find $\phi_1 \otimes_R S_1 : S_1 \to S_1 \otimes_R S_1$; composing with the multiplication map gives us the identity map, so the rank of ϕ_1 must be 1. Therefore, the map $\phi_1 \otimes_R S_2$ also has rank 1 over S_2; in particular $S_1 \otimes_R S_2$ is non-zero.

We wish to show that α is a full map with respect to ρ. In order to do this, we need to show that the minimal rank with respect to ρ of an f.g. projective submodule of $\begin{pmatrix} S_1 & S_1 \otimes_R S_2 \\ 0 & 0 \end{pmatrix}$ containing the image of α is $\frac{1}{2}$. So, we need to know what the submodules of $\begin{pmatrix} S_1 & S_1 & S_2 \\ 0 & 0 \end{pmatrix}$ look like. We leave it as an easy exercise for the reader to show that such a submodule M takes the form $\begin{pmatrix} eS_1 & eS_1 \otimes_R S_2 \\ 0 & 0 \end{pmatrix} \oplus \begin{pmatrix} 0 & M \\ 0 & 0 \end{pmatrix}$ where M is an S_2-complement of $eS_1 \otimes_R S_i$ in $S_1 \otimes_R S_2$.

Consequently, its rank is equal to $\frac{1}{2}\rho_1(eS_1) + \frac{1}{2}\rho_2(M)$ where ρ_i is the rank of S_i modules over S_i.

Suppose that there is some submodule $\begin{pmatrix} eS_1 & eS_1 \otimes_R S_2 \\ 0 & 0 \end{pmatrix} \oplus \begin{pmatrix} 0 & M \\ 0 & 0 \end{pmatrix}$ of $\begin{pmatrix} S_1 & S_1 \otimes_R S_2 \\ 0 & 0 \end{pmatrix}$ containing the image of α such that it has rank q less than $\frac{1}{2}$. Then consider the map $\gamma : R \to (1-e)S_1$ which sends 1 to $1-e$; over S_2 we find that the rank of $\gamma \otimes_R S_2 : S_2 \to (1-e)S_1 \otimes_R S_2$ is at most $2(q - \frac{1}{2}\rho_1(eS_1))$, since the image of S_2 lies in the image of M under left multiplication by $(1.e)$. So, the rank of $\gamma \otimes_R S_1 : S_1 \to (1-e)S_1 \otimes_R S_1$ is equal to the rank of $\gamma \otimes_R S_2$ which is at most $2(q - \frac{1}{2}\rho_1(eS_1))$.

That is, the image of S_1 under $\gamma \otimes_R S_1$ in $(1-e)S_1 \otimes_R S_1$ lies in some submodule M_1 of rank over S_1 at most $2(q - \frac{1}{2}\rho_1(eS_1))$; therefore,

$\begin{pmatrix} 0 & 1\otimes_R S_1 \\ 0 & 0 \end{pmatrix}$ lies in $\begin{pmatrix} eS_1 & eS_1\otimes_R S_1 \\ 0 & 0 \end{pmatrix} \oplus \begin{pmatrix} 0 & M_1 \\ 0 & 0 \end{pmatrix}$, inside the ring

$\begin{pmatrix} S_1 & S_1\otimes_R S_1 \\ 0 & S_1 \end{pmatrix}$. This shows that the rank of the map defined by left

multplication by $\begin{pmatrix} 0 & 1\otimes_R 1 \\ 0 & 0 \end{pmatrix}$ from $\begin{pmatrix} 0 & 0 \\ 0 & S \end{pmatrix}$ to $\begin{pmatrix} S_1 & S_1\otimes_R S_1 \\ 0 & 0 \end{pmatrix}$ with

respect to the rank function that assigns these modules the rank ½ is at most q, since its image lies in $\begin{pmatrix} eS_1 & eS_1\otimes_R S_1 \\ 0 & 0 \end{pmatrix} \oplus \begin{pmatrix} 0 & M_1 \\ 0 & 0 \end{pmatrix}$ whose rank

with respect to this rank function is at most $\tfrac{1}{2}\rho_1(eS_1) + q - \tfrac{1}{2}\rho_1(eS_1) = q$. However, it must be full, since it becomes invertible under the homomorphism to $M_2(S_1)$ given by $\begin{pmatrix} S_1 & S_1\otimes_R S_1 \\ 0 & S_1 \end{pmatrix} \to \begin{pmatrix} S_1 & S_1 \\ S_1 & S_1 \end{pmatrix}$ where the map from $S_1\otimes_R S_1$ to S_1 is the multiplication map.

Therefore, we have a contradiction if we assume that α is not a full map. So, $T_\alpha \cong M_2(S_1 \underset{R}{\sqcup} S_2)$ has a unique rank function by 5.2, and the universal localisation of $S_1 \underset{R}{\sqcup} S_2$ is a simple artinian ring, which we call the <u>simple artinian coproduct</u> of S_1 and S_2 <u>amalgamating</u> R, and write as $S_1 \underset{R}{\circ} S_2$.

<u>Universal localisation and Sylvester rank functions</u>

Our method for constructing homomorphisms from a ring to suitable simple artinian rings will be to adjoin the universal inverses to some set of maps between f.g. projectives; provided there are enough maps of a suitable sort the ring we shall obtain will be local with simple artinian residue ring. In order to prove these results, we shall need to investigate ways of extending Sylvester module and map rank functions from a ring to a universal localisation of the ring. This theory allows us to prove Cohn's results on homomorphisms from a ring to skew fields which we shall need in preparation for the general case.

<u>Theorem 7.4</u> Let R be a ring with a Sylvester rank function, ρ; let Σ be a collection of maps between f.g. projectives over R whose rank is equal to the rank of the identity map on the domain and codomain. Then, the universal localisation of R at Σ, R_Σ, does not vanish, and the Sylvester map rank function on R extends to a Sylvester map rank function on R, ρ_Σ,

that takes values in the subgroup of the reals that ρ does.

Proof: Once we show how to extend ρ to R_Σ, and show that this extension is well-defined, it follows that R_Σ does not vanish.

First, we note that we can assume that Σ is upper multiplicatively closed, since the rank of all elements of the upper multiplicative closure of Σ is equal to the rank of the domain and codomain.

As we have done in similar situations, we define the extension of ρ to R_Σ by Cramer's rule, and then, we show that it is well-defined by Malcolmson's criterion.

Let $\beta : R_\Sigma \otimes_R P \to R_\Sigma \otimes_R Q$ be a map between f.g. induced projective modules over R_Σ. Then, by Cramer's rule, there is an equation:

$$(\alpha\ \alpha_1)\begin{pmatrix} I_{P'} & \beta_1 \\ 0 & \beta \end{pmatrix} = (\alpha\ \alpha')$$

We attempt to define $\rho_\Sigma(\beta)$ by $\rho_\Sigma(\beta) = \rho(\alpha\ \alpha') - \rho(P')$. Suppose that we have two equations of the above form:

$$(\alpha\ \alpha_1)\begin{pmatrix} I_{P'} & \beta_1 \\ 0 & \beta \end{pmatrix} = (\alpha\ \alpha')\ ,\ (\gamma\ \gamma_1)\begin{pmatrix} I_{P''} & \beta_2 \\ 0 & \beta \end{pmatrix} = (\gamma\ \gamma')$$

Then,

$$\left(\begin{array}{ccc|c} \alpha & \alpha_1 & 0 & 0 \\ 0 & \gamma_1 & \gamma & \gamma_1 \end{array}\right)\left(\begin{array}{c|c} & \beta_1 \\ I & \beta \\ & \beta_2 \\ \hline 0 & 0 \end{array}\right) = \left(\begin{array}{ccc|c} \alpha & \alpha_1 & 0 & \alpha' \\ 0 & \gamma_1 & \gamma & \gamma' \end{array}\right)$$

so, by Malcolmson's criterion, there is an equation

$$\left(\begin{array}{cccccc|c} \alpha & \alpha_1 & 0 & 0 & 0 & 0 & \alpha' \\ 0 & \gamma_1 & \gamma & \gamma_1 & 0 & 0 & \gamma' \\ 0 & 0 & 0 & 0 & \delta_1 & 0 & 0 \\ 0 & 0 & 0 & 0 & 0 & \delta_2 & \varepsilon_2 \\ \hline 0 & 0 & 0 & I & \varepsilon_1 & 0 & 0 \end{array}\right) = \begin{pmatrix} \mu_1 \\ \mu_2 \\ \mu_3 \\ \mu_4 \\ \sigma \end{pmatrix}(\nu|\tau)$$

From this, we construct two equations:

$$1.\quad \left(\begin{array}{cccccc|c} \alpha & \alpha_1 & 0 & 0 & 0 & 0 & 0 \\ 0 & \gamma_1 & \gamma & \gamma_1 & 0 & 0 & \gamma' \\ 0 & 0 & 0 & 0 & \delta_1 & 0 & 0 \\ 0 & 0 & 0 & 0 & 0 & \delta_2 & 0 \\ \hline 0 & 0 & 0 & I & \varepsilon_1 & 0 & 0 \end{array}\right) = \left(\begin{array}{c|c} \mu_1 & 0 \\ \mu_2 & \gamma' \\ \mu_3 & 0 \\ \mu_4 & 0 \\ \hline \sigma & 0 \end{array}\right)\left(\begin{array}{c|c} \nu & 0 \\ \hline 0 & I \end{array}\right)$$

and

$$2.\quad \left(\begin{array}{cccccc|c} \alpha & \alpha_1 & 0 & 0 & 0 & 0 & -\alpha' \\ 0 & \gamma_1 & \gamma & \gamma_1 & 0 & 0 & 0 \\ 0 & 0 & 0 & 0 & \delta_1 & 0 & 0 \\ 0 & 0 & 0 & 0 & 0_1 & \delta_2 & -\varepsilon_2 \\ \hline 0 & 0 & 0 & I & \varepsilon_1 & 0 & 0 \end{array}\right) = \left(\begin{array}{c|c} \mu_1 & 0 \\ \mu_2 & \gamma' \\ \mu_3 & 0 \\ \mu_4 & 0 \\ \hline \sigma & 0 \end{array}\right)\left(\begin{array}{c|c} \nu & -\tau \\ \hline 0 & I \end{array}\right)$$

Note that if χ_1 lies in S, then $\rho\begin{pmatrix}\chi_1 & 0 \\ \phi & \chi_2\end{pmatrix} = \rho(\chi_1) + \rho(\chi_2)$, for arbitrary ϕ, since by axiom 4 for a Sylvester map rank function, it is at least that, whilst

$$\begin{pmatrix}\chi_1 & 0 \\ \phi & \chi_2\end{pmatrix} = \begin{pmatrix}\chi_1 & 0 \\ 0 & I\end{pmatrix}\begin{pmatrix}I & 0 \\ \phi & I\end{pmatrix}\begin{pmatrix}I & 0 \\ 0 & \chi_2\end{pmatrix}$$

so that axiom 2 shows that it cannot be greater. A similar argument applies to show that $\rho\begin{pmatrix}\chi_1 & \phi' \\ 0 & \chi_2\end{pmatrix} = \rho(\chi_1) + \rho(\chi_2)$ for arbitrary ϕ'. Since $\begin{pmatrix}\chi_1\chi_2 & 0 \\ 0 & I\end{pmatrix}$ is associated to $\begin{pmatrix}\chi_2 & I \\ 0 & \chi_1\end{pmatrix}$, we deduce that $\rho(\chi_1\chi_2) = \rho(\chi_2)$ whenever $\chi_1\chi_2$ is defined, and $\chi_1 \in \Sigma$.

So, $\rho(\text{LHS1}) = \rho(\delta_1) + \rho(\delta_2) + \rho(P) + \rho(\alpha\alpha_1) + \rho(\gamma\gamma')$; whilst $\rho(\text{LHS2}) = \rho(\delta_1) + \rho(\delta_2) + \rho(P) + \rho(\gamma\gamma_1) + \rho(\alpha\alpha')$.

But $\rho(\text{RHS1}) = \rho(\text{RHS2})$, since both $\begin{pmatrix}\nu & 0 \\ 0 & I\end{pmatrix}$ and $\begin{pmatrix}\nu & -\tau \\ 0 & I\end{pmatrix}$ lie in Σ; so, $\rho(\text{RHS1}) = \rho(\text{LHS1}) = \rho(\text{LHS2}) = \rho(\text{RHS2})$. So, $\rho(\alpha\alpha') - \rho(\alpha\alpha_1)$ is equal to $\rho(\gamma\gamma') - \rho(\gamma\gamma_1)$, from which it is clear that ρ is well-defined. We have to be careful what we mean at this point since this rank function is not as yet defined on association classes of maps over R_Σ but only up to multiplication by invertible maps over R on the left. We shall however show that our rank function defined on maps between induced f.g. projectives over R_Σ satisfies axioms 1 to 4 for a Sylvester rank function from which it follows that it is well-defined on association classes of maps over R_Σ. It follows that it extends to a Sylvester map rank function since it restricts to a Sylvester matrix rank function and so it extends to a Sylvester map rank function by lemma 7.2.

Axioms 1 and 3 are clear, so we are left with 2 and 4 which are rather less obvious.

First, we show that $\rho_\Sigma\begin{pmatrix} I_Q & \alpha_1 \\ O & \alpha \end{pmatrix} = \rho_\Sigma(\alpha) + \rho(Q)$, for arbitrary α_1.

Suppose that we have equations:

$$(\beta|\beta_1)\left(\begin{array}{c|c} I_{Q_1} & \alpha' \\ \hline O & \alpha \end{array}\right) = (\beta|\beta') \ , \ (\gamma|\gamma_1)\left(\begin{array}{c|c} I_{Q_2} & \alpha'' \\ \hline O & \alpha_1 \end{array}\right) = (\gamma|\gamma')$$

then,

$$\left(\begin{array}{cc|cc} \gamma & \gamma_1 & O & O \\ \hline O & O & \beta & \beta_1 \end{array}\right)\left(\begin{array}{cc|c|c} I_{Q_2} & O & & \alpha'' \\ O & I_{Q_3} & O & \alpha_1 \\ \hline & & I_{Q_1} & \alpha' \\ \cline{3-4} & O & O & \alpha \end{array}\right) = \left(\begin{array}{cc|c|c} \gamma & \gamma_1 & O & \gamma' \\ \hline O & O & \beta & \beta' \end{array}\right)$$

We note that $\begin{pmatrix} I_{Q_2} & O & O & \alpha'' \\ O & I_{Q_3} & O & \alpha_1 \\ O & O & I_{Q_1} & \alpha' \\ O & O & O & \alpha \end{pmatrix}$ is associated to $\begin{pmatrix} I_{Q_2} & O & O & \alpha'' \\ O & I_{Q_1} & O & \alpha' \\ O & O & I_{Q_3} & \alpha_1 \\ O & O & O & \alpha \end{pmatrix}$

so that $\rho_\Sigma\begin{pmatrix} I_{Q_3} & \alpha_1 \\ O & \alpha \end{pmatrix} = \rho\begin{pmatrix} \gamma & \gamma_1 & O & \gamma' \\ O & O & \beta & \beta' \end{pmatrix} - \rho(Q_1 \oplus Q_2) = \rho(\gamma\gamma_1) + \rho(\beta\beta') - \rho(Q_1 \oplus Q_2)$

since $(\gamma\gamma_1)$ is in Σ. This equals $\rho(Q_3) + \rho_\Sigma(\alpha)$ as we stated.

Next, we show that $\rho_\Sigma(\alpha\beta) \le \rho_\Sigma(\alpha), \rho_\Sigma(\beta)$ when α is an induced map. Also, we shall show that if α is in Σ, then $\rho_\Sigma(\alpha\beta) = \rho_\Sigma(\beta)$.

Let $\alpha:P_0 \to P_1$ be an induced map over R_Σ, and let $\beta:P_1 \to P_2$ be a map between induced projectives over R_Σ. Then $\rho_\Sigma(\alpha\beta) + \rho(P_1)$ $= \rho_\Sigma\begin{pmatrix} I_{P_1} & \beta \\ 0 & \alpha\beta \end{pmatrix} = \rho_\Sigma\begin{pmatrix} I_{P_1} & \beta \\ -\alpha & 0 \end{pmatrix} = \rho_\Sigma\begin{pmatrix} \beta & -I_{P_1} \\ 0 & \alpha \end{pmatrix}$, where the last equations arise from multiplication on the left by invertible maps over R.

If $(\gamma\ \gamma_1)\begin{pmatrix} I_{P'} & \beta_1 \\ 0 & \beta \end{pmatrix} = (\gamma\ \gamma')$ then

$$\begin{pmatrix} \gamma & \gamma_1 & 0 \\ 0 & 0 & I_{P_0} \end{pmatrix}\begin{pmatrix} I_{P'} & \beta_1 & 0 \\ 0 & \beta & -I \\ 0 & 0 & \alpha \end{pmatrix} = \begin{pmatrix} \gamma & \gamma' & -\gamma_1 \\ 0 & 0 & \alpha \end{pmatrix} = \begin{pmatrix} I_{Q'} & 0 \\ 0 & \alpha \end{pmatrix}\begin{pmatrix} I_{Q'} & -\gamma_1 \\ 0 & I_{P_1} \end{pmatrix}\begin{pmatrix} \gamma & \gamma' & 0 \\ & 0 & I_{P_1} \end{pmatrix} \text{where } Q'$$

is the domain of $(\gamma\gamma_1)$.

So, $\rho_\Sigma\begin{pmatrix} \beta & -I \\ 0 & \alpha \end{pmatrix} = \rho\begin{pmatrix} \gamma & \gamma' & -\gamma_1 \\ 0 & 0 & \alpha \end{pmatrix} - \rho(P') \le \rho(\alpha) + \rho(Q') - \rho(P')$,

$\rho(\gamma\gamma') + \rho(P_1) - \rho(P')$. Since $(\gamma\gamma_1)$ is in Σ, $\rho(Q') = \rho(\gamma\gamma_1) = \rho(P') + \rho(P_1)$. Therefore, we see that $\rho(\alpha) + \rho(Q') - \rho(P') = \rho(\alpha) + \rho(P_1)$.

So, $\rho_\Sigma(\alpha\beta) + \rho(P_1) = \rho_\Sigma\begin{pmatrix} \beta & -I \\ 0 & \alpha \end{pmatrix} \le \rho_\Sigma(\alpha) + \rho(P_1),\ \rho_\Sigma(\beta) + \rho(P_1)$ as we want.

If we assume α is in Σ,

$$\rho_\Sigma\begin{pmatrix} \beta & -I \\ 0 & \alpha \end{pmatrix} = \rho\begin{pmatrix} \gamma & \gamma' & -\gamma_1 \\ 0 & 0 & \alpha \end{pmatrix} - \rho(P') = \rho(\gamma\gamma') + \rho(\alpha) - \rho(P')$$

$= \rho(\alpha) + \rho_\Sigma(\beta) = \rho(P_1) + \rho_\Sigma(\beta)$. Hence, $\rho_\Sigma(\alpha\beta) = \rho_\Sigma\begin{pmatrix} \beta & -I \\ 0 & \alpha \end{pmatrix} - \rho(P_1) = \rho_\Sigma(\beta)$.

In general, suppose that we have maps over R between induced f.g. projectives $\alpha:P_0 \to P_1$, and $\beta:P_1 \to P_2$. Suppose that we have an equation:

$$(\gamma\ \gamma_1)\begin{pmatrix} I_P & \alpha_1 \\ 0 & \alpha \end{pmatrix} = (\gamma\ \gamma')$$

then $\rho(I_P) + \rho_\Sigma(\alpha\beta) = \rho_\Sigma\begin{pmatrix} I_P & \alpha_1\beta \\ 0 & \alpha\beta \end{pmatrix} = \rho_\Sigma(\gamma\ \gamma')\begin{pmatrix} I_P & \alpha_1\beta \\ 0 & \alpha\beta \end{pmatrix} = \rho_\Sigma(\gamma\ \gamma')\begin{pmatrix} I_P & 0 \\ 0 & \beta \end{pmatrix}$

$\le \rho_\Sigma(\gamma\gamma')$, $\rho(P) + \rho_\Sigma(\beta)$ so $\rho_\Sigma(\alpha\beta) \le \rho(\gamma\gamma') - \rho(P) = \rho_\Sigma(\alpha)$, and $\rho_\Sigma(\alpha\beta) \le \rho_\Sigma(\beta)$ which is axiom 2.

Let α_1, α_2 and β be maps over R_Σ between induced f.g. projectives such that $\begin{pmatrix} \alpha_1 & 0 \\ \beta & \alpha_2 \end{pmatrix}$ is defined. Suppose that we have two equations:

$$(\gamma\ \gamma_1)\begin{pmatrix} I_{P_1} & \alpha' \\ 0 & \alpha_1 \end{pmatrix} = (\gamma\ \gamma')\ ,\ (\delta\ \delta_1)\begin{pmatrix} I_{P_2} & \alpha'' \\ 0 & \alpha_2 \end{pmatrix} = (\delta\ \delta')$$

Then

$$\rho_\Sigma\begin{pmatrix} \alpha_1 & 0 \\ \beta & \alpha_2 \end{pmatrix} + \rho(P_1 \oplus P_2) = \rho_\Sigma\begin{pmatrix} I & & 0 & \alpha'' \\ & & \alpha' & 0 \\ 0 & 0 & \alpha_1 & 0 \\ 0 & 0 & \beta & \alpha_2 \end{pmatrix} = \rho_\Sigma\left(\begin{pmatrix} \delta & 0 & 0 & \delta_1 \\ 0 & \gamma & \gamma_1 & 0 \end{pmatrix}\begin{pmatrix} I & & 0 & \alpha \\ & & \alpha' & 0 \\ 0 & 0 & \alpha_1 & 0 \\ 0 & 0 & \beta & \alpha_2 \end{pmatrix}\right)$$

$$= \rho_\Sigma\begin{pmatrix} \delta & 0 & \delta_1\beta & \delta' \\ 0 & \gamma & \gamma' & 0 \end{pmatrix} = \rho_\Sigma\begin{pmatrix} \gamma & \gamma' & 0 & 0 \\ 0 & \delta_1\beta & \delta & \delta' \end{pmatrix}$$ so that the problem reduces to the case where α_1, α_2 are induced maps.

So, suppose that we have an equation:

$$(\varepsilon\ \varepsilon_1)\begin{pmatrix} I_P & \beta_1 \\ 0 & \beta \end{pmatrix} = (\varepsilon\ \varepsilon')$$

then $$\rho_\Sigma\begin{pmatrix} \alpha_1 & 0 \\ \beta & \alpha_2 \end{pmatrix} + \rho(P) = \rho_\Sigma\begin{pmatrix} I_P & \beta_1 & 0 \\ 0 & \alpha_1 & 0 \\ 0 & \beta & \alpha_2 \end{pmatrix} = \rho_\Sigma\left(\begin{pmatrix} \varepsilon & 0 & \varepsilon_1 \\ 0 & I & 0 \end{pmatrix}\begin{pmatrix} I_P & 0 & 0 \\ 0 & \alpha_1 & 0 \\ 0 & \beta & \alpha_2 \end{pmatrix}\right)$$

$$= \rho\begin{pmatrix} \varepsilon & \varepsilon' & \varepsilon_1\alpha_2 \\ 0 & \alpha_1 & 0 \end{pmatrix} = \rho\begin{pmatrix} \alpha_1 & 0 & 0 \\ \varepsilon' & \varepsilon & \varepsilon_1\alpha_2 \end{pmatrix} = \rho\left(\begin{array}{c|cc} \alpha_1 & 0 & 0 \\ \hline \varepsilon' & (\varepsilon\ \varepsilon_1) & \begin{smallmatrix} I_P & 0 \\ 0 & \alpha_2 \end{smallmatrix} \end{array}\right) \geq \rho(\alpha_1) + \rho(\alpha_2) + \rho(I_P)$$

which proves axiom 4.

We have shown that our rank function defined on maps between induced f.g. projectives satisfies the axioms of a Sylvester map rank function. It follows that it extends uniquely to a Sylvester map rank function as we have outlined before.

In the case where the Sylvester map rank function on R takes values in $\mathbb{Z}$, we can say a great deal about the universal localisation of R at all the maps between f.g. projectives whose rank is equal to the rank of the identity map on the domain and codomain. In fact, we obtain Cohn's classification of homomorphisms to skew fields.

<u>Theorem 7.5</u> Let R be a ring with a Sylvester rank function ρ taking values in $\mathbb{Z}$; then the universal localisation of R at Σ, the collection

of those maps between f.g. projectives whose rank is equal to the rank of the domain and codomain is a local ring whose residue ring is a skew field. The kernel of the map from R to the residue ring consists of those elements which define maps of rank 0. Therefore, the equivalence classes of maps from R to skew fields are in 1 to 1 correspondence with the Sylvester rank functions taking values in $\mathbb{Z}$, or equivalently, with the prime matrix ideals.

Proof: By the last theorem, we know that R_Σ exists and ρ induces a Sylvester rank function on R_Σ taking values in $\mathbb{Z}$. Let I be the subset of R_Σ consisting of those elements whose rank is 0; then it is an additive subgroup because $a + b = (1\ 1)\begin{pmatrix} a & 0 \\ 0 & b \end{pmatrix}\begin{pmatrix} 1 \\ 1 \end{pmatrix}$ and it is closed under multiplication so it is an ideal.

If x is not in I, its rank is 1. Let $(\alpha\ \alpha_1)\begin{matrix} I & \beta \\ 0 & x \end{matrix} = (\alpha\alpha')$ be some equation given by Cramer's rule; then because x has rank 1, $(\alpha\ \alpha')$ must be in Σ. So, x is invertible.

So, as we stated R_Σ is a local ring with maximal ideal I, and R_Σ/I is a skew field. Clearly, the Sylvester map rank function induced by the map $R \to R_\Sigma/I$ is the one we began with. Conversely, if we start with a homomorphism from R to a skew field, $R \to F$, inducing a Sylvester map rank function ρ, there is a map from R_Σ to F extending the map from R to F, and the kernel of this map is I, for if a is in I, and a has non-zero image in F, we consider some equation given by Cramer's rule: $(\alpha\ \alpha_1)\begin{pmatrix} I & \beta \\ 0 & x \end{pmatrix} = (\alpha\ \alpha')$. Since a has non-zero image in F, the left hand side is invertible over F and so must be the right hand side; however, the right hand side is not invertible, since its rank given by the map rank function induced by the homomorphism from R to F is not equal to the rank of the identity map on its domain and codomain. We have a contradiction, so I lies in the kernel of the map from R_Σ to F. Therefore, R_Σ/I embeds in F as the skew subfield of F generated by the image of R.

The remark on the kernel of the map from R to F is trivial, since the map from R to F induces the Sylvester rank function ρ.

Ring coproducts and rank functions

In this section, we shall show how Sylvester module and map rank functions on a k-algebra R may be extended to Sylvester module and map rank functions on the ring $M_n(k) \sqcup_k R$. This will allow us to show that every Sylvester module rank function on a k-algebra arises from a homomorphism to a simple artinian ring by showing that the Sylvester module rank function taking values in $\frac{1}{n}\mathbb{Z}$ extends to a Sylvester module rank function on $M_n(k) \sqcup_k R$ also taking values in $\frac{1}{n}\mathbb{Z}$; $M_n(k) \sqcup_k R$ is isomorphic to $M_n(R')$ for some R' and by Morita equivalence there is a Sylvester module rank function on R' taking values in $\mathbb{Z}$; we have shown that this is induced by a homomorphism to a skew field, F, and this shows that the original rank function must be induced by the homomorphisms: $R \to M_n(k) \sqcup_k R \tilde{\to} M_n(R') \to M_n(F)$.

It remains to find a way of extending the Sylvester map rank function on R to one on $M_n(k) \sqcup_k R$. The idea is to mimic what we know the result would have to be if there were a homomorphism from R to a simple artinian ring S inducing our Sylvester map rank function. In this case, we have homomorphisms: $M_n(k) \sqcup_k R \to M_n(k) \sqcup_k S \to M_n(k) \circ_k S$, which induce a Sylvester map rank function on $M_n(k) \sqcup_k R$ extending that on R. Given a map $\alpha: P \to Q$ between f.g. projective modules over $M_n(k) \sqcup_k R$, the rank of $\alpha \otimes (M_n(k) \circ_k S)$ is the minimal rank of an f.g. projective module over $M_n(k) \sqcup_k S$ containing the image of $\alpha \otimes (M_n(k) \sqcup_k S)$. It is hoped that this discussion will help to motivate the definition we shall propose later on for the map rank function on $M_n(k) \sqcup_k R$.

Unfortunately, before we can begin, we need to go through a certain amount of technical work on coproducts over a skew field. Rather than referring the reader to an earlier chapter, we shall reproduce some of the definitions here; also, we shall prove one of the coproduct theorems here, since it is precisely the technical details of the proof that we need to examine.

Let R_0 be a skew field and $\{R_\lambda : \lambda \cup \Lambda\}$ a family of R_0-rings. Set $M = \Lambda \cup \{0\}$. We form the ring coproduct $R = \sqcup_{R_0} R_\lambda$ and consider an induced module which has the form $N = \bigoplus_\mu N_\mu \otimes_{R_\mu} R$. For each λ, we choose a right basis over R_0 of the form $\{1\} \cup T$ for the ring R_λ, and for each μ, we choose a basis over R_0, S_μ, for N_μ. Write $S = \bigcup_\mu S_\mu$, and $T = \bigcup_\lambda T_\lambda$.

If $t \in T_\lambda$, it is associated to λ; if $S \in S_\lambda$, it is associated to λ; if $s \in S_0$, it is associated to no index. A monomial is an element of S or a formal product $st_1t_2 \ldots t_n$, $s \in S$ and $t_i \in T$ such that no two successive terms are associated to the same index. Let U be the set of monomials; an element of U is associated to λ if and only if its last factor (in S or

T) is associated to λ. Every element of U is associated to some index except for the elements of S_0. We denote by $U_{\sim\lambda}$ those elements that are not associated to λ. We recall again without proof theorem 2.4.

Theorem 2.4 Let all terms be as above. Then a right basis for N over R_0 is the set U. For each λ, N is the direct sum as R_λ module of N_λ and a free R_λ module on the basis $U_{\sim\lambda}$.

Given λ and $u \in U_{\sim\lambda}$, we denote by $c_{\lambda u}:N \to R_\lambda$ the R_λ linear right 'coefficient of u' map given by the decomposition of the theorem. For $u \in U$, we denote by $c_{0u}:N \to R_0$ the R_0 linear 'right coefficient of u' map given by the decomposition of N as R_0 module in the theorem. For $\lambda \in \Lambda$, the λ-support of an element x of N is the finite set of monomials in $U_{\sim\lambda}$ such that $c_{\lambda u}(x)$ is not 0; x has empty λ-support if and only if it lies in N_λ. The 0-support (or support) of an element x consists of those monomials such that $c_{0u}(x)$ is not 0.

The degree of a monomial $st_1\ldots t_n$ is $(n + 1)$, and the degree of an element s of S is 1. The degree of an element of N is the maximal degree of an element in its support. We define an element x in N to be λ-pure if all the monomials in its support of maximal degree are associated to the index λ. It is 0-pure if and only if it is not λ-pure for any index in Λ.

We well-order the sets S and T in some way, and then we well-order U by degree and then lexicographically reading from left to right. Next, we well-order M, making 0 the least element. We well-order $M \times U$ first by the degree of the second factor, and afterwards lexicographically from left to right. Let H be the set of almost everywhere zero functions from $M \times U$ to $\mathbb{N}$ well-ordered lexicographically reading from highest to lowest in $M \times U$.

Given any element x in N, its leading term is the largest element in its support; its λ-leading term is the maximal element in its λ-support (if it has any).

Given a homomorphism of f.g. induced modules $\alpha:\bigoplus_\mu M_\mu \otimes_{R_\mu} R \to N$, we wish to find an isomorphism $\beta: \bigoplus_\mu M'_\mu \otimes_{R_\mu} R \to \bigoplus_\mu M_\mu \otimes_{R_\mu} R$ of induced modules such that the image of α is isomorphic to $\bigoplus_\mu (\alpha\beta)M'_\mu \otimes_{R_\mu} R$. It turns out that we may find such an isomorphism composed of transvections and free transfer maps. In order to present the proof, we introduce following Bergman the notion of a well-positioned family of submodules of N.

A family of R_μ submodules of N, $\{L_\mu\}$ is said to be well-positioned if and only if the following conditions are satisfied:

A_μ: $\forall \mu \in \mathcal{M}$, all elements of L_μ are μ-pure;

$B_{\mu_1\mu_2}$: the μ_1-support of L_{μ_1} contains no monomial u which is also the μ_1-leading term of some non-μ_1-pure element xa, x in L_{μ_2}, a in R, and if $\mu_1 = \mu_2$, $\deg xa > \deg x$.

It is not hard to show that if $\{L_\mu\}$ is a well-positioned family $\Sigma_\mu L_\mu R$ is naturally isomorphic to $\oplus_\mu L_\mu \otimes_{R_\mu} R$. The idea for the construction of the isomorphism is that if $\{\alpha(M_\mu): \mu \in \mathcal{M}\}$ is not a well-positioned family, we find some free transfer or transvection $\beta_1: \oplus_\mu M'_\mu \otimes_{R_\mu} R \to \oplus_\mu M_\mu \otimes_{R_\mu} R$ so that $\{\alpha\beta_1(M'_\mu): \mu \in \mathcal{M}\}$ is a 'better positioned' family; by using the well-ordering, we can make sure that this process terminates and so, it gives us the isomorphism that we were looking for.

<u>Theorem 7.6</u> Let $\alpha: \oplus M_\mu \otimes_{R_\mu} R \to N$ be a homomorphism of f.g. induced modules, where N is as described above. Then, there is an isomorphism of induced modules $\beta: \oplus_\mu M'_\mu \otimes_{R_\mu} R \to \oplus_\mu M_\mu \otimes_{R_\mu} R$ which is a finite composition of free transfers and transvections such that $\{\alpha\beta(M'_\mu): \mu \in \mathcal{M}\}$ is a well-positioned family of submodules of N.

Proof: We associate to the map α a function $h_\alpha: \mathcal{M} \times U \to \{0,1\}$ by

$$h(\mu,u) = \begin{cases} 1, & \text{if } u \text{ is in the } \mu\text{-support of } \alpha(M_\mu); \\ 0, & \text{otherwise.} \end{cases}$$

This is an almost everywhere zero function since M_μ is f.g. and is non-trivial for only finitely many μ.

Suppose that $\alpha(M_{\lambda_1})$ is not λ_1-pure for some λ_1; then there is some $\alpha(x)$ in $\alpha(M)$ with λ_1-leading term u such that $c_{\lambda_1 u}(x) = 1$; the map $c: M \to R_{\lambda_1}$ is an R_{λ_1} split surjection; so, $M_{\lambda_1} \cong \ker c_{\lambda_1} \oplus xR_{\lambda_1}$ where xR_{λ_1} is free of rank 1. We perform the free transfer between $\oplus_\mu M_\mu \otimes_R R$ and $\oplus M'_\mu \otimes_{R_\mu} R$, where $M'_\lambda = M_\lambda$ for $\lambda \neq \lambda_1$; $M_0 = M_0 \oplus \tilde{x}R_0$; $M'_{\lambda_1} = \ker c_{\lambda_1 u}$. It is the identity map on M'_λ for $\lambda \neq \lambda_1$; it maps M'_0 to M_0, maps $\ker c_{\lambda_1 u}$ to $\ker c_{\lambda_1 u}$ and sends $\tilde{x}$ to x. It is clear that if β_1 is this free transfer, $h_{\alpha\beta_1}: \mathcal{M} \times U \to \{0,1\}$ is a smaller function in the well-ordering of such functions that we described earlier.

If, however, $\alpha(M_O)$ is not O-pure, there is some element $\alpha(x)$ in $\alpha(M_O)$ that is λ_1-pure with leading term u such that $c_{Ou}(x) = 1$; $M_O \cong \ker c_{Ou} \oplus xR_O$; so, we perform the free transfer $\beta_1: \bigoplus_\mu M'_\mu \otimes_{R_\mu} R \to M_\mu \otimes_{R_\mu} R$, where $M'_\lambda = M_\lambda$, $\lambda \neq \lambda_1$; $M_O = \ker c_{Ou}$; $M'_{\lambda_1} = M_{\lambda_1} \oplus \tilde{x}R_{\lambda_1}$. β_1 has the obvious effect. Again, it is clear that $h_{\alpha\beta_1}$ is less than h_α under these conditions. Therefore, after a suitable finite sequence of free transfers, we may assume that each $\alpha(M_\mu)$ is μ-pure.

Next, suppose that for some pair μ_1, μ_2, $B_{\mu_1\mu_2}$ fails for the family of modules $\{\alpha(M_\mu)\}$; that is, there is an element $\alpha(x)$ in $\alpha(M_{\mu_1})$ such that its μ_1-support contains a monomial u that is the μ_1-leading term of a non-μ_1-pure element $\alpha(y)a$ where y is in M_{μ_2}, a is in R, and if $\mu_1 = \mu_2$, $\deg(ya) > \deg(y)$. We may assume that $c_{\mu_1 u}(\alpha(y)a) = 1$.

We have a functional $M_{\mu_1} \to R_{\mu_1}$ given by $c_{\mu_1 u}$ which we extend to a functional on $M = \bigoplus_\mu M_\mu \otimes_{R_\mu} R$ in the way we described when describing transvections. Left multiplication of this functional by ya now gives us an endomorphism t of M of square 0; so $\beta = I_M - t$ is an automorphism of M which is a transvection. It is an easy check to show that $h_{\alpha\beta}$ is a smaller function than h_α; therefore, after a suitable finite sequence of transvections and (possibly more) free transfers, $\beta = \prod_i \beta_i$ we may ensure that $\{\alpha\beta(M'_\mu) : \mu \in \mathcal{M}\}$ is a well-positioned family of submodules.

Now we shall show that a well-positioned family justifies its name.

<u>Theorem 7.7</u> Let $\{M_\mu : \mu \in \mathcal{M}\}$ be a well-positioned family of submodules in N; then $\Sigma M_\mu R \cong \bigoplus_\mu M_\mu \otimes_R R$ in the natural way.

Proof: Given $\mu \in \mathcal{M}$, we choose for each monomial u that is the leading term of some element of M an element q in M_μ having this leading term with co-efficient 1; we denote the set of such q's by Q_μ. It is clear from the well-ordering that each Q_μ is an R_O-basis of M_μ. For each $\lambda \in \Lambda$, and monomial u that is a λ-leading term of an element of M_O, we choose an element q in M_O that has this term as λ-leading term with co-efficient 1, and we denote the set of such q's by $Q_{O\lambda}$. Every element of M_O has a λ-leading term so $Q_{O\lambda}$ is an R_O-basis of M_O for each λ.

The elements of Q_λ are associated to λ, and the elements of $Q_{O\lambda}$ are associated to all indices in $\Lambda - \{\lambda\}$.

We consider the set V consisting of all elements of the form $qt_1 \ldots t_n$, where q lies in $\bigcup_\lambda (Q_\lambda \cup Q_{O\lambda})$ and no two successive terms in the above monomial are associated to the same index, together with $\bigcup_\mu Q_\mu$. We shall show that these elements in N have distinct leading terms, and so, are rightly R_O-linearly independent. Therefore, they form an R_O basis of $\Sigma_\mu M_\mu R$, from which our theorem is clear.

By the well-ordering of U, the leading term of $qt_1 \ldots t_n$ is $ut_1 \ldots t_n$, where q lies in Q_λ and u is the leading term of q, or where q lies in $Q_{O\lambda}$ and u is the λ-leading term of q. So, we consider an equality of the form $ut_1 \ldots t_m = u't'_1 \ldots t'_n$ where $m \geq n$; u comes from q in M_{μ_2}, u' from q' in M_{μ_1}. Assume $m = n$; we obtain by cancellation, $u = u'$. If $q \in Q_\lambda$, u is associated to λ, so, $q' \in Q_O$, for q' cannot lie in Q_λ by construction. Therefore, q' is in Q_O or $Q_{O\lambda'}$ for $\lambda' \neq \lambda$. If q' is in $Q_{O\lambda'}$ the support of $q' \in M_O$ contains a monomial u' which is the leading term of a pure element q in M_λ, which contradicts $B_{O\lambda}$. If q' is in Q_O, $m = n = 0$, and we obtain as above a contradiction to $B_{O\lambda}$. The same applies if $q \in Q_O$.

If q is in $Q_{O\lambda}$, then q' cannot lie in Q_λ since $m = n = 0$; so q is in $Q_{\lambda'}$, $\lambda \neq \lambda'$. Once more, we have a contradiction to $B_{O\lambda'}$.

So, $m > n$, and, by cancellation, we find $ut_1 \ldots t_{m-n} = u'$.

Since q' is in M_{μ_1}, we see that if $\mu_1 \neq 0$, t_{m-n} is in T_{μ_1} so the μ_1-support of q' in M_{μ_1} contains $ut_1 \ldots t_{m-n}$ (or u if $m = n + 1$). This also is the μ_1-leading term of the non-μ_1-pure element $qt_1 \ldots t_{m-n}$ (or q if $m = n + 1$), which contradicts $B_{\mu_1\mu_2}$. If $\mu_1 = 0$, then since the support of q' in M_O contains $ut_1 \ldots t_{m-n}$ the leading term of the pure element $qt_1 \ldots t_{m-n}$ we again have a contradiction to $B_{\mu_1\mu_2}$.

This completes the proof, since we see that the elements of V are independent and span $\Sigma_\mu M_\mu R$.

Bergman also uses a direct way of finding well-positioned families of submodules that generate a given submodule of an induced module.

Thus, let $L \subseteq N$; we define L_μ to be the R_μ submodule of L consisting of the elements whose μ-support does not contain the μ-leading term of some non-μ-pure element of L. By construction, the family $\{L_\mu\}$ is well-positioned; it is not hard to show that $L = \Sigma_\mu L_\mu R$, which as we have just seen is naturally isomorphic to $\oplus L_\mu \otimes_{R_\mu} R$. In the case where L is a f.g. submodule, we may regard L as being generated by some finite family F of f.g. R_μ submodules of L, $\{M_\mu\} = F$; so, $\Sigma_\mu M_\mu R = L$. To such a family, we associate a function $h_F: \mathcal{M} \times U \to \{0,1\}$ by

$$h_F(\mu,u) = \begin{cases} 1 & \text{if } u \text{ is in the } \mu\text{-support of some element of } M_\mu; \\ 0 & \text{otherwise.} \end{cases}$$

We shall characterise the family of submodules $\{L_\mu\}$ where L_μ is the set of elements whose μ-support does not contain the μ-leading term of some non-μ-pure element of L by the property that the associated function is minimal in the well-ordering of such functions.

<u>Theorem 7.8</u> Let L be an f.g. submodule of N, where N is the module we have considered throughout this chapter. Let L_μ be the R_μ submodule consisting of those elements whose μ-support does not contain the μ-leading term of some non-μ-pure element of L. Then, the function h_F associated to the family of submodules $F = \{L_\mu\}$ is minimal over all possible finite families of submodules that generate L.

Proof: Assume that we have a family with smaller associated function than h_F. Given a family of submodules $F_1 = \{M_\mu\}$ that generate L, we have a map $\alpha: \oplus M_\mu \otimes_{R_\mu} R \to N$ whose image is L.

The proof of the last theorem showed that the method used there of passing to a map $\alpha': \oplus_\mu M_\mu \otimes_{R_\mu} R \to N$ such that the family $F_2 = \{\alpha'(M_\mu)\}$ is well-positioned always forces the function h_{F_2} to be less than h_{F_1}. So, we may begin by assuming that our family $\{\alpha(M_\mu)\}$ is well-positioned. Therefore, $L = \oplus M_\mu \otimes_{R_\mu} R$ in the natural way.

Suppose that the μ_1-support of M_{μ_1} contains the μ_1-leading term u of some non-μ_1-pure element x in L. We assume that $c_{\mu_1 u}(x) = 1$. We note that the degree of x is less than the degree of an element of M_{μ_1}

whose μ_1-support contains u.

We wish to show that $L = \Sigma M'_\mu R$ where $M'_\mu = M_\mu$ $\mu \neq \mu$, whilst $M'_{\mu_1} = \beta(M_{\mu_1})$ where $\beta(m) = m - xc_{\mu_1 u}(m)$. So, β is the identity on elements of degree less than equal to x.

Clearly, if $F' = \{M'_\mu : \mu \in M\}$, $h_{F'} < h_{F'}$ so, if $\Sigma M'_\mu R = L$, the well-ordering of our functions imply that after finitely many steps (which may include more operations that change F' into a well-positioned family) we reach a family $F = \{M_\mu\}$ where no element of M_μ contains the μ-leading term of some non-μ-pure element of L, and $\Sigma M_\mu R = L$. Clearly, $M_\mu = L_\mu$. Since our associated functions have decreased at each step, this will prove our theorem.

We show that x lies in $\Sigma M_\mu R$ from which it follows that this must equal L. We use the notation of 7.7. We look at an expression of x with respect to the R_O basis V. No elements of degree greater than that of x can occur since all elements of V have distinct leading terms. So, all elements in this expression of x must lie in $\Sigma M_\mu R$, since their term from Q either lies in $M_\mu = M'_\mu$, for $\mu \neq \mu_1$, or else it is fixed by β; and so must x.

We can begin the proof that every Sylvester map rank function on a k-algebra R taking values in $\frac{1}{n}\mathbb{Z}$ extends to a Sylvester map rank function on $R' = M_n(k) \sqcup_k R$ that also takes values in $\frac{1}{n}\mathbb{Z}$. Set $R_O = k$; $R_1 = M_n(k)$; and $R_2 = R$.

The first point to notice about a Sylvester map rank function is that the rank of a map $\alpha: P \to Q$ depends only on the image of α in Q; for, if $\alpha': P' \to Q$ has the same image then α factors through α' and α' factors through α, so they have the same rank. So, our problem is to assign to a given f.g. submodule of an f.g. R' submodule of an f.g. projective R' module a rank so that the associated map rank function is Sylvester.

Let $P \cong \oplus_\mu P_\mu \otimes_{R_\mu} R$ be an f.g. R' module, where we may assume that P_O is O; we identify P with the module N that we have been discussing in this chapter, and so, we identify N_μ with P_μ. So, we have bases Q_μ of each P_μ over k, bases $T_1 \cup \{1\}$ of $M_n(k)$ over k, and $T_2 \cup \{1\}$ of R over k, and consequently, a basis U of P over k consisting of monomials of the form q or $qt_1 \ldots t_n$. We also have the well-orderings previously defined of U and of the functions from $\{O,1,2\} \times U$ to the natural numbers.

Let $L \subseteq P$ be some f.g. R' submodule of P; let L_μ be the R_μ submodule of elements of L whose μ-support does not contain the μ-leading term of some non-μ-pure element. For $\mu = 0,1$, L_μ has a rank as R_μ module, where ρ_μ are the usual rank functions on k and $M_n(k)$. For $\mu = 2$, the map rank function allows us to define the rank of the inclusion of L_2 in P as a map of R_2 modules (recall that we earlier how to extend a Sylvester map rank function in a canonical way to give us the rank of maps from an f.g. module to an arbitrary module). We define the pre-rank of L to be $\rho_0(L_0) + \rho_1(L_1) + \rho_2(L_2 \subseteq P)$. The rank of L is defined to be the minimal possible pre-rank of an f.g. submodule of P that contains L. So, the rank of a map $\alpha:\rho' \to P$ between f.g. projective R' modules is defined by the formula: $\bar{\rho}(\alpha) = \min_{L \supseteq \text{in } \alpha} \{\text{pre-rank } (L)\}$. We shall show that $\bar{\rho}$ is a Sylvester map rank function.

If we have a map $\alpha: \oplus_\mu P'_\mu \otimes_{R_\mu} R \to P$, we could assign what we might regard as a pre-pre-rank, which is $\rho_0(\alpha(P'_0)) + \rho_1(\alpha(Q'_1)) + \rho_2(\alpha|_{P'_2})$; our next lemma, which shows that the pre-rank is the minimal possible 'pre-pre-rank' as we consider the composition of α with isomorphisms of $\oplus_\mu Q'_\mu \otimes_{R_\mu} R$ with other induced modules is the main step of the proof that $\bar{\rho}$ is a Sylvester map rank function.

<u>Lemma 7.9</u> Let $L \subseteq P$ be some f.g. R' submodule of P; let $\{M_\mu : \mu = 0,1,2\}$ be R_μ submodules of L such that $\sum_\mu M_\mu R' = L$; then the pre-rank of $L \subseteq P$ is the minimal value of $\rho_0(M_0) + \rho_1(M_1) + \rho_2(M_2 \subseteq P)$.

Proof: Theorems 7.6 and 7.8 give us a finite sequence of operations that pass from any given trio of modules $\{M_\mu\}$ that generate L to the trio $\{L_\mu\}$ where L_μ is the set of elements whose μ-support does not contain the μ-leading term of some non-μ-pure element of L. So, we simply need to show that these operations cannot increase the 'pre-pre-rank'.

So, suppose that M_0 is not 0-pure; then there exists x in M_0, where x is 1- or 2-pure, its leading term is u, and $c_{0u}(x) = 1$; if x is 1 pure, we replace:

M_0 by $\ker c_{0u}|M_0$;

M_1 by $M_1 + xR_1$;

M_2 by M_2.

The pre-pre-rank does not increase.

If x is 2-pure, we replace

$$M_0 \text{ by } \ker{}_{c0u}|M_0;$$

$$M_1 \text{ by } M_1;$$

$$M_2 \text{ by } M_2 + xR_2$$

and again the pre-pre-rank cannot increase, since the contribution of the 0-term has decreased by 1, whilst the 2-term has increased by at most 1.

If M_1 is not 1-pure, we simply transfer a free R_1 module to an R_0 free module so the pre-pre-rank cannot increase.

If M_2 is not 2 pure, there is an element x in M_2 with 2-leading term u such that $c_{2u}(x) = 1$; in this case $M_2 = \ker c_{2u} + xR_2$ where xR_2 is a free direct summand of M_2; it is also, however, a free direct summand of P, since $c_{2u}(x) = 1$, so,
$\rho_2(M_2 \subseteq P) = \rho_2(\ker(c_{2u}|M_2) \subseteq P) + 1.$

In this case, our operation replaces:

$$M_0 \text{ by } M_0 + xR_0;$$

$$M_1 \text{ by } M_1;$$

$$M_2 \text{ by } \ker(c_{2u}|M_2)$$

and by our previous remarks, we see that the pre-pre-rank cannot increase.

Thus, we have shown that our efforts to make each M_μ μ-pure do not increase the pre-pre-rank. We consider next the transvections we need to make them well-positioned.

If M_{μ_1} contains the μ_1-leading term u for some non-μ_1-pure element xa for x in M_{μ_2}, a in R' and if $\mu_1 = \mu_2$, $\deg xa > \deg x$, then for $\mu_1 = 0,1$, we know that the transvection takes M_{μ_1} to a homomorphic image of itself whilst fixing the other M_μ. If, however, $\mu_1 = 2$, more care is required. We assume that $c_{2u}(xa) = 1$. We define a functional c_{2u} on M_2 by $f: M_2 \subseteq P \to R_2$; the transvection fixes M_0 and M_1 and sends M_2 to M_2' the image of $L_{M_2} - xaf$; this map has as a left factor the inclusion of M_2 in P; consequently, by axiom 2 for a Sylvester map rank function it follows that $\rho_2(M'_2 \subseteq P)$ is at most $\rho_2(M_2 \subseteq P)$ and the pre-pre-rank does not increase.

Finally, we consider the case where $\{M_\mu : \mu \in \mathcal{M}\}$ is a well-positioned family of submodules of L such that $\sum_\mu M_\mu R' = L$ but M_{μ_1} for some $\mu_1 = 0,1,2$ contains in its μ_1-support some monomial u that is the μ_1-leading term of some non-μ_1-pure element x in L; we also may assume that $c_{\mu_1 u}(x) = 1$.

If $\mu_1 = 0,1$, we see that the new M_{μ_1} is a homomorphic image of M_{μ_1}, whilst the remaining M_μ do not change. If $\mu_1 = 2$, we fix M_0 and M_1 and send M_2 to the image of $M_2 \subseteq P \xrightarrow{Ip-x\,C_{2u}} P$, where $I_P - xc_{2u}$ is clearly an R_2 linear map; again $M_2 \subseteq P$ is a left factor so the pre-pre-rank cannot increase, which completes the proof of the lemma, since there are no more operations that we have to worry about.

<u>Theorem 7.10</u> Let $R_2 = R$ be a k-algebra with a Sylvester rank function ρ taking values in $\frac{1}{n}\mathbb{Z}$; let $R_0 = k$, and let $R_1 = M_n(k)$ with the standard rank functions ρ_0, ρ_1 respectively; let $R' = R_1 \underset{R_0}{\sqcup} R_2$; let $\bar{\rho}$ be the rank function on maps between f.g. projectives over R' defined in the foregoing. Then, $\bar{\rho}$ is a Sylvester map rank function taking values in $\frac{1}{n}\mathbb{Z}$. Consequently, there is a homomorphism from R to a simple artinian ring inducing the Sylvester map rank function ρ.

Proof: We need to show that axioms 1 to 4 for a Sylvester map rank function actually hold for $\bar{\rho}$.

1 is clear.

We consider 2. Let $\alpha : P_1 \to P_2$, $\beta : P_2 \to P_3$ be a pair of maps where we have an expression of P_i as an induced module up our sleeve when we need it. The image of $\alpha\beta$ lies in the image of β, and so, $\bar{\rho}(\alpha\beta) \le \bar{\rho}(\beta)$. $\bar{\rho}(\alpha)$ is the pre-rank of some f.g. submodule containing im α $L = \bigoplus_\mu L_\mu \otimes_{R_\mu} R' \subseteq P_2$ where L_μ is the set of elements whose μ-support does not contain the μ-leading term of some non-μ-pure element of L. That is, $\bar{\rho}(\alpha) = \rho_0(L_0) + \rho_1(L_1) + \rho_2(L_2 \subseteq P)$; the image of $\alpha\beta$ lies in $\sum_\mu (L_\mu)R'$, so, by the last lemma,

$$\bar{\rho}(\alpha\beta) \le \rho_0(\beta(L_0)) + \rho_1(\beta(L_1)) + \rho_2(\beta(L_2) \subseteq P).$$

The map $L_2 \subseteq P_2 \xrightarrow{\beta} P_3$ has $L_2 \subseteq P_2$ as left factor so

$$\bar{\rho}(\alpha\beta) \le \rho_0(L_0) + \rho_1(L_1) + \rho_2(L_2 \subseteq P_2) = \bar{\rho}(\alpha).$$

So, 2 holds.

For the time being, we remark only that $\rho\begin{pmatrix}\alpha & 0\\ 0 & \beta\end{pmatrix} \le \rho(\alpha) + \rho(\beta)$; the rest of 3 follows from 4.

The reason that a Sylvester map rank function works is essentially because of axiom 4; so, one would expect this to be the most difficult part of the proof. It is.

Consider a map $\begin{pmatrix}\alpha & 0\\ \gamma & \beta\end{pmatrix}:\begin{pmatrix}P\\ \oplus\\ P'\end{pmatrix} \to \begin{pmatrix}Q\\ \oplus\\ Q'\end{pmatrix}$.

Let L be an f.g. submodule of $Q \oplus Q'$ that contains the image of $\begin{pmatrix}\alpha & 0\\ \gamma & \beta\end{pmatrix}$ such that $\bar{\rho}\begin{pmatrix}\alpha & 0\\ \gamma & \beta\end{pmatrix}$ is the pre-rank of L. We consider the action of the projection $p:Q \oplus Q' \to Q$ on L; the idea is to find R_μ submodules of L, $\{L'_\mu\}$ such that $\{p(L'_\mu)\}$ are a well-positioned family, which implies that $p(L) \cong \oplus_\mu p'(L'_\mu) \otimes_{R_\mu} R$; also, we wish the pre-rank of L to equal $\rho_0(L'_0) + \rho_1(L'_1) + \rho_2(L'_2 \subseteq Q \oplus Q')$. This allows us to express $\rho_2(L'_2 \subseteq Q \oplus Q')$ as being $\rho_2\begin{pmatrix}\alpha_2 & 0\\ \gamma_2 & \beta_2\end{pmatrix}$ for a map $\begin{pmatrix}\alpha_2 & 0\\ \gamma_2 & \beta_2\end{pmatrix}:\begin{pmatrix}P''\\ \oplus'\\ P''_2\end{pmatrix} \to \begin{pmatrix}Q\\ \oplus\\ Q'\end{pmatrix}$ where P''_i are f.g. R_2 projectives, $(\ker p|L'_0)R' + (\ker p|L'_1)R' + (\operatorname{im}\beta_2)R' \supseteq \operatorname{im}\beta$, and also, $(\operatorname{im} p|L'_0)R' + (\operatorname{im} p|L'_i)R' + (\operatorname{im}\alpha_2)R' \supseteq \operatorname{im}\alpha$. From this, 4 follows as a corollary of 4 for ρ_2 on R_2.

Let L_μ be the set of elements of L whose μ-support does not contain the μ-leading term of some non-μ-pure element of L: then, we know that $L \cong \oplus_\mu L_\mu \otimes_{R_\mu} R'$ in the natural way, and the pre-rank of L is equal to

$\rho_0(L_0) + \rho_1(L_1) + \rho_2(L_2 \subseteq Q \oplus Q')$.

Let $p:Q \oplus Q' \to Q$ be the projection on Q; if $\{\rho(L_\mu)\}$ is not a well-positioned family, we perform a series of free transfers and transvections on the family of modules $\{L_\mu\}$ until their image do form a well-positioned family. It is clear that the free transfers do not increase the pre-pre-rank of the family of modules $\{L_\mu\}$ at any stage, so, we are left to worry about the transvections. These arise when the image $p(L_\mu)$ are all μ-pure but some $p(L_\mu)$ contains the μ-leading term u of a non-μ-pure element xa chosen so that $c_{\mu u}(xa) = 1$, x is in $p(L_{\mu'})$ and if $\mu = \mu'$, $\deg(xa) > \deg(x)$. It is clear that the only case that worries us occurs when $\mu = 2$. In this case, we alter L_2 to the image of

$L_2 \subset Q \oplus Q' \xrightarrow{I - x'a(pc_{2u})} Q \oplus Q'$, where x' is a pre-image of x under p. Since this map factors through $L_2 \subseteq Q \oplus Q'$ it cannot give a greater pre-pre-rank. So, eventually, we reach a family of modules $\{L'_\mu\}$ such that the pre-pre-rank associated to $\{L'_\mu\}$ is at most that of the family $\{L_\mu\}$ we began with and the images $P(L'_\mu)$ form a well-positioned family. Since the modules $\{L'_\mu\}$ generate L, their pre-pre-rank must be at least that of the family $\{L_\mu\}$ by lemma 7.9. Since $p(L'_\mu)$ form a well-positioned family, $p(L) \cong \bigoplus_\mu p(L'_\mu) \otimes_{R_\mu} R'$. So, the kernel of p restricted to L is $\bigoplus_\mu \ker p|L'_\mu \otimes_{R_\mu} R'$. We have exact sequences $0 \to L'_\mu \cap Q' \to L'_\mu \to p(L'_\mu) \to 0$ for $\mu = 0,1,2$.

From this, we wish to deduce that

$\rho_\mu(L'_\mu) = \rho_\mu(L'_\mu \cap Q') + \rho_\mu(p(L'))$, $\mu = 0,1$, which is clear; and also, that $\rho_2(L'_2 \subseteq Q \oplus Q') \geq \rho_2(L'_2 \cap Q' \subseteq Q') + \rho_2(p(L'_2) \subseteq Q)$, which needs an argument. In fact, it is not obvious that we can define $\rho_2((L'_2 \cap Q') \subset Q')$ because $L'_2 \cap Q'$ need not be f.g. Instead we approximate it by f.g. modules that are good enough.

Let $\alpha_2 : P''_1 \to Q$ be some R_2 linear map from an f.g. R_2 projective module whose image equals $p(L'_2)$; then, we have a map from P to L'_2 lifting α_2, we write its composite with $L'_2 \subseteq Q \oplus Q'$ as $\begin{pmatrix}\alpha_2\\ \gamma_2\end{pmatrix} : P \to \begin{pmatrix}Q\\ \oplus\\ Q'\end{pmatrix}$

L'_2 is a f.g.; therefore, we can find a f.g. R_2 submodule of $L'_2 \cap Q'$, M_2 such that together with the image of $\begin{pmatrix}\alpha_2\\ \gamma_2\end{pmatrix}P$ it generates the whole of L'_2;

we may further assume that $(L'_0 \cap Q')R' + (L_1 \cap Q')R' + M_2R')$ contains the image of β.

We find some R_2 linear map $\rho_2 : P''_2 \to Q'$ whose image is M_2, and we construct the composite map: $\begin{pmatrix}\alpha_2 & 0\\ \gamma_2 & \beta_2\end{pmatrix} : \begin{pmatrix}P''_1\\ \oplus\\ P''_2\end{pmatrix} \to \begin{pmatrix}Q\\ \oplus\\ Q'\end{pmatrix}$

By construction, the image of $\begin{pmatrix}\alpha_2 & 0\\ \gamma_2 & \beta_2\end{pmatrix}$ is L'_2 so that

$\rho_2(L_2' \subseteq Q \oplus Q')$ is $\rho_2\begin{pmatrix}\alpha_2 & 0\\ \gamma_2 & \beta_2\end{pmatrix} \geq \rho_2(\alpha_2) + \rho_2(\beta_2)$.

Since $\operatorname{im} \beta \subseteq (L_0' \cap Q')R' + (L_1' \cap Q')R' + (\operatorname{im} \beta_2)R'$, $\bar{\rho}(\beta) \leq \rho_0(L' \cap Q') + \rho_1(L_1' \cap Q') + \rho_2(\beta_2)$, and similarly, $\bar{\rho}(\alpha) \leq \rho_0(p(L')) + \rho_1(p(L')) + \rho_2(\alpha_2)$.

So, $\bar{\rho}\begin{pmatrix}\alpha & 0\\ \gamma & \beta\end{pmatrix}$ = pre-rank of $L = \rho_0(L_0') + \rho_1(L_1') + \rho_2(L_2' \subseteq Q \oplus Q')$

which is $\rho_0(L_0') + \rho_1(L') + \rho_2\begin{pmatrix}\alpha_2 & 0\\ \gamma_2 & \beta_2\end{pmatrix}$, which is greater than or equal to

$\rho(L_0' \cap Q') + \rho_0(p(L_0')) + \rho_1(L_1' \cap Q') + \rho_1(p(L_1')) + \rho_2(\alpha_2) + \rho_2(\beta_3)$, which

is at least $\bar{\rho}(\alpha) + \bar{\rho}(\beta)$. This is axiom 4, and also completes the proof of 3.

We have extended ρ_2 on $R_2 = R$ to a Sylvester map rank function $\bar{\rho}$ on $M_n(k) \sqcup_k R$, and it is clear that $\bar{\rho}$ takes values in $\frac{1}{n}\mathbb{Z}$. $M_n(k) \sqcup_k R \cong M_n(A)$ where A is the centraliser of the first factor of the coproduct. So, by Morita equivalence, $\bar{\rho}$ induces a Sylvester map rank function on A that takes values in $\mathbb{Z}$. By theorem 7.5, the universal localisation of A at the set of maps between f.g. projectives whose rank is equal to the rank of the identity map on the domain and the codomain is a local ring with a skew field F for residue class ring. Moreover, the map $A \to F$ induces the rank function on A. Consequently, the composite map $R \to M_n(k) \sqcup_k R \tilde{\to} M_n(A) \to M_n(F)$ induces the rank function ρ_2 on R. As we saw in theorem 7.5, the kernel of the map consists of the elements r in R such that $\rho(R/rR) = 1$, or equivalently, such that $\rho(r) = 0$.

<u>Theorem 7.11</u> A Sylvester rank function on a k-algebra R taking values in $\frac{1}{n}\mathbb{Z}$ arises from a homomorphism to a simple artinian ring, $M_n(D)$, where D is a skew field. The kernel of the map consists of the elements of rank 0.

We summarise the contents of this chapter so far with regard to homomorphisms from k-algebras to simple artinian rings.

<u>Theorem 7.12</u> Let R be a k-algebra; then there is a 1 to 1 correspondence between equivalence classes of homomorphisms from R to simple artinian rings of the form $M_n(D)$, where D is a skew field, and Sylvester rank functions that take values in $\frac{1}{n}\mathbb{Z}$.

This result has the following interesting consequence

Theorem 7.13 Every right artinian k-algebra embeds in a simple artinian ring.

Proof: Call the right artinian ring A. Let s be the length of A as a module over itself. We define a module rank function by $\rho(M) = (\text{length of } M)/s$. Clearly, this is a Sylvester module rank function. By theorem 7.11, there is a homomorphism from A to a simple artinian ring S, which induces this rank function on A, and the kernel of this homomorphism is the set of elements a in A such that $\rho(A/aA) = \rho(A)$; the only element for which this is true is 0.

In order to find the modification necessary to extend theorem 7.12 to a theorem about homomorphisms from an arbitrary ring to simple artinian rings, it is useful to see what can go wrong in general. We look at possible Sylvester map rank functions on artinian rings. If A is an artinian ring, then $G_0(R) \cong \mathbb{Z}^k$, where t is the number of non-isomorphic simple modules. We assign the simple modules ranks in the non-negative rationals, and it is clear that this defines a rank function on $G_0(A)$ that induces a Sylvester module rank function on A, and so, a Sylvester map rank function on A provided that we normalise it to ensure that the rank of the free module of rank 1 has rank 1. In the case where A is a k-algebra, we have just shown that these must all arise from a homomorphism to a simple artinian ring, and it is easy to check that they arise from the representations given by the f.g. projective left A modules. However, for a general artinian ring, it is easy to see that they do not all arise from a homomorphism to a simple artinian ring. For example, consider the ring $\mathbb{Z}/p^2$, which has the unique simple module $\mathbb{Z}/p$, to which we assign the rank ½. It has no embedding in a simple artinian ring, but this rank function is faithful. What has gone wrong in this example is that we have the rank ½ associated to multiplication by p on the free module of rank 1, whereas in a simple artinian ring, the rank of an integer must be 0 or 1, depending on whether it is invertible or 0. This turns out to be the only modification we need to our previous theory.

Theorem 7.14 Let R be a ring. Then the equivalence classes of homomorphisms from R to simple artinian rings are in 1 to 1 correspondence with the Sylvester map rank functions that take values in $\frac{1}{n}\mathbb{Z}$ such that the rank of an integer is 0 or 1.

Proof: All that is left to do to prove this is to show that every Sylvester map rank function of the right type arise from homomorphisms to simple artinian rings. Let ρ be such a rank function on the ring R.

Assume that $\rho(n) = 0$ for a non-zero integer n. Let I be the set of elements in R such that $\rho(r) = 0$. We have already remarked that this forms an ideal of R. Moreover, if mn lies in I for integers m and n, one of m and n must lie in I for if $\rho(m) = 1$, $\rho(mn) = \rho(n)$. Therefore, the intersection of I with $\mathbb{Z}$ is a prime ideal of $\mathbb{Z}$. Consider R/I, which is a $\mathbb{Z}/p$ algebra for some prime p. We define a Sylvester map rank function on R/I by $\bar{\rho}(\alpha) = \rho(\alpha)$ where $\bar{\alpha}$ is the image of α in R/I; it is simple to check that $\bar{\rho}$ is well-defined. Consequently, it is induced by some homomorphism $R/I \to S$ to a simple artinian ring. The map from R to S induces ρ on R clearly.

Conversely, assume that $\rho(n) = 1$ for all non-zero integers n. Then, by 7.4, the rank function ρ on R extends to a Sylvester map rank function on $R_{\mathbb{Z}^*}$ since all the elements in $\mathbb{Z}^*$ have the same rank as the identity map on their domain and codomain. But $R_{\mathbb{Z}^*}$ is a $\mathbb{Q}$-algebra, so the rank function on it is induced by a homomorphism to a simple artinian ring, S. The induced map from R to S induces on R.

Maximal epic subrings and dominions in simple artinian rings

One of the difficulties we encountered at the beginning of our study of a homomorphism from R to a simple artinian ring S was that the maximal epic R-ring in S was not simple artinian. In this section, we shall study an individual homomorphism from a ring to a simple artinian ring more closely; we shall be interested in the maximal epic R-ring and the dominion of R in S. Our principal results state that they must both be semiprimary rings.

We had better define the terms in the last paragraph. Given a homomorphism from a ring A to a ring B, we consider the set of all epic A-rings in B; since the ring generated by two epic A-rings is an epic A-ring, and the union of epic A-rings is an epic A-ring, there is a unique <u>maximal epic A-ring</u> in B. The <u>dominion</u> of A in B is the maximal subring D of B such that homomorphisms from B that agree on the image of A in

B must agree on D; it is clear that the maximal epic A-ring in B lies in the dominion of A in B. The dominion has a simple description given in the next lemma. First, we define a useful construction. If R is a ring and M is an R bimodule, the trivial extension of R by M is the ring whose R bimodule structure is $R \oplus M$, and whose multiplication is defined by $M^2 = 0$.

<u>Lemma 7.15</u> Let $\phi : A \to B$ be a ring homomorphism; then the dominion of A in B is the centraliser of $1 \otimes 1$ in the B,B bimodule $B \otimes_A B$.

Proof: Certainly, the dominion centralises $1 \otimes 1$, for we have the following two ring homomorphisms from B to the trivial extension of B by $B \otimes_A B$; the first is the identity map from B to B; the second sends $b \in B$ to $b + b \otimes 1 - 1 \otimes b$. They agree on A, so, they must agree on D, which shows that $d \otimes 1 = 1 \otimes d$ for d in D.

Conversely, if we have two homomorphisms from B to a ring C that agree on A, ϕ_1 and ϕ_2, the B,B bimodule $\phi_1(B)\phi_2(B)$ is a quotient of $B \otimes_A B$, so that the centraliser of $1 \otimes 1$ lies in D.

In order to give the reader some better idea of the dominion, we prove the following interesting lemma. The <u>double centraliser</u> of a subring A of a ring B is the ring of elements of B that centralise the centraliser of A in B.

<u>Lemma 7.16</u> Let $\phi : A \to B$ be a ring homomorphism. The dominion of A in B lies in the double centraliser of A in B. If A is a k-subalgebra of $M_n(k)$, the dominion of A in $M_n(k)$ is its double centraliser.

Proof: Let c lie in the centraliser of A in B; we define two homomorphisms from B to $B[x : x^2 = 0]$ one of which is the identity map, the other of which sends b in B to $b + x(bc - cb)$. It is clear that they agree on the image of A, and so, they agree on D the dominion of A in D; consequently, D lies in the double centraliser of A in B.

In the case where $B \cong M_n(k)$ and A is a k-subalgebra, we may prove the converse. We use the last lemma. D is the centraliser of $1 \otimes 1$ in $M_n(k) \otimes_A M_n(k)$. As $M_n(k)$, $M_n(k)$ bimodule, $M_n(k) \otimes_A M_n(k) \cong {}^tM_n(k)$ for some integer t. Let $c_1, \ldots c_t$ be the images of $1 \otimes 1$ in the direct summands on the right of this isomorphism; then the dominion of A in $M_n(k)$

is the centraliser of these elements, and $c_1, \ldots c_t$ all lies in the centraliser of A in $M_n(k)$.

We wish to study the dominion of a ring A under a homomorphism from A to a ring B; our next lemma gives us a method of attack.

Lemma 7.17 Let $\phi: A \to B$ be a ring homomorphism; then the dominion of A in B is the endomorphism ring of the module M over the ring $\begin{pmatrix} B & B\otimes_A B \\ O & B \end{pmatrix}$

where M is given by the exact sequence: $O \to (O \quad B) \overset{\alpha}{\to} (B \quad B\otimes_A B) \to M \to Q$
$\alpha(O \quad b) = (O \quad 1\otimes b)$.

Proof: All endomorphisms of M lift to endomorphisms of $(B \quad B\otimes_A B)$ that normalise $(O \quad 1\otimes_A B)$. The endomorphism ring of $(B \quad B\otimes_A B)$ is defined by left multiplication by elements of B; so we look for elements t of B such that $t\otimes 1 = 1\otimes t$. Hence, our lemma follows.

If B is semiprimary, so is $\begin{pmatrix} B & B\otimes_A B \\ O & B \end{pmatrix}$; therefore, we wish to show that the endomorphism ring of a finitely presented module over a semiprimary ring is itself semiprimary. This has been shown by Bjork (71), who also had results in the direction of the corresponding theorem for left perfect rings; we shall present a proof of both these results next, generalising Pjork's theorem.

Theorem 7.18 Let R be a left perfect ring (or a semiprimary ring), and let M be a finitely presented right module over R; then the endomorphism ring of M over R is left perfect (or semiprimary respectively).

Proof: First, we see that because M is finitely presented it has the descending chain condition on submodules with a bounded number of generators, since this is true for its projective cover by the left perfect condition. Consequently, it is the direct sum of finitely many indecomposables.

Let P be the projective cover of M; so there is an exact sequence: $O \to I \to P \to M \to O$ where I is finitely generated. Any endomorphism of M lifts to an endomorphism α of P such that $\alpha(I) \subseteq I$. We intend to show that every endomorphism of M lifts to a nilpotent endomorphism of P or acts invertibly on a direct summand of M.

Let $\bar{\alpha}:M \to M$ lift to $\alpha:P \to P$, $\alpha(I) \subseteq I$. There exists an integer n such that $\alpha^n P = \alpha^{n+1} P$ and $\bar{\alpha}^n M = \bar{\alpha}^{n+1} M$. Since $\alpha:\alpha^n P \to \alpha^n P$ is surjective, and $End_R(P)$ is left perfect, $\alpha^n P = eP$ for some idempotent endomorphism of P, e. If $\alpha^n P = 0$, then $\bar{\alpha}^n M = 0$, since $(1-e)P$ cannot map onto M as P is the projective cover of M. $\bar{\alpha}^n M \cong (\alpha^n P + I)/I = eP/I \cap eP = eP/eI$, which shows that $\bar{\alpha}^n M$ is finitely presented; further, $\bar{\alpha}:\bar{\alpha}^n M \to \bar{\alpha}^n M$ is a surjective map; we shall show that this forces it to be an isomorphism on $\bar{\alpha}^n M$. Let J/eI be the kernel of this map; so, $eP/J \cong eP/eI$; so, there exists an endomorphism $\beta:eP \to eP$ such that $\beta J = eI$. Since $I \subseteq radP$, $eI \subseteq rad(eP)$ and eP is the projective cover of $\bar{\alpha}^n M$; therefore, β must be an invertible map since it is inducing an isomorphism module the radical of P; if $\beta J = eI \subsetneq J$, we obtain an infinite descending chain of modules of bounded number of generators, $J \supsetneq \beta J \supsetneq \ldots \supsetneq \beta^n J \ldots$, which is impossible, so $J = eI$, and α is an isomorphism on $\alpha^n M$ with inverse γ. We see that $\gamma^n \alpha^n$ gives a projection from M onto $\bar{\alpha}^n M$, which proves our dichotomy.

Next, we show that if M is an indecomposable module, its endomorphism ring is left perfect (or semiprimary if R is semiprimary). Since M is indecomposable, every endomorphism on M is invertible or else it lifts to a nilpotent endomorphism of the projective cover P of M as we have just shown. First, we deal with the left perfect case. Let $\bar{\alpha}_i$ be elements of $End_R(M)$ that lift to nilpotent endomorphisms α_i on P. The elements $\bar{\alpha}_i$ are themselves nilpotent and so, $\bar{\alpha}_1 \ldots \bar{\alpha}_n M \subsetneq \bar{\alpha}_1 \ldots \bar{\alpha}_{n-1} M$; hence, $\bar{\alpha}_1 \ldots \bar{\alpha}_n P + I \subsetneq \bar{\alpha}_1 \ldots \bar{\alpha}_{n-1} P + I$ provided that $\alpha_1 \ldots \alpha_{n-1} P \not\subseteq I$. The number of generators of $\alpha_1 \ldots \alpha_n P + I$ is bounded for all n, so eventually, $\alpha_1 \ldots \alpha_n P \subset I$, which implies that $\bar{\alpha}_1 \ldots \bar{\alpha}_n = 0$. This shows that the radical of $End_R(M)$ is left T-nilpotent and so $End_R(M)$ is left perfect by Tachikawa (73). This leaves the case where R is semiprimary. Then $End_R(P)$ is semiprimary, so that there exists an integer N such that $\alpha^N = 0$ for any nilpotent endomorphism of P; hence, if $\bar{\alpha}$ is an endomorphism of M that is not invertible $\bar{\alpha}^N = 0$. By the Nagata, Higman theorem, the radical of $End_R(M)$ is nilpotent, so $End_R(M)$ is semiprimary.

If M is an arbitrary f.g. module, we write $M \cong \bigoplus_{i=1}^{n} M_i$ where M_i are indecomposable; by theorem 2.4 of Tachikawa (73) $End_R(M)$ is left perfect if R is, and it is semiprimary if R is semiprimary.

Putting this theorem together with the preliminary discussion, we deduce what we were after.

Theorem 7.19 Let $\phi : R \to S$ be a homomorphism from a ring R to a semiprimary ring S; then the dominion of R in S is also semiprimary.

Proof: By lemma 7.17, the dominion is the endomorphism ring of a suitable finitely presented module over a semiprimary ring; so it is semiprimary by theorem 7.18.

This theorem has a number of interesting consequences but we begin with the following lemma.

Lemma 7.20 Let $\{D_i\}$ be a family of subrings of a ring all of which are their own dominion in R; then $\bigcap_i D_i$ is its own dominion in R.

Proof: The dominion of $\bigcap_i D_i$ in R is the centraliser of $1 \otimes 1$ in $R \otimes_{\cap D_i} R$. By considering the natural map from $R \otimes_{\cap D_i} R$ to $R \otimes_{D_i} R$, we see that this centraliser must lie in D_i and so in $\cap D_i$; therefore $\cap D_i$ is its own dominion.

A semisimple artinian ring is always its own dominion, so there is the following result.

Theorem 7.21 The intersection of a collection of semisimple artinian subrings of a ring is always semiprimary.

The kernel of a derivation $d : R \to M$ is always its own dominion, as one sees by considering ring homomorphisms from R to the trivial extension of R by M; similarly, it is clear that the fixed points of an endomorphism of a ring R is a dominion. It follows that the kernel of a family of derivations or the fixed points of a family of ring endomorphisms on a semiprimary ring is also semiprimary. Similar results are true for perfect rings. The following corollary will be needed in the next section.

Corollary 7.22 Let S be semiprimary with subring R; then the centraliser of R is semiprimary.

Proof: The centraliser is the kernel of the inner derivations determined by the elements of R; so we can apply the foregoing discussion.

We should like to study the maximal epic R-subring of a ring R under a homomorphism from R to a semiprimary ring, S. First of all, the maximal epic R-subring must lie in the dominion of R in S, and this suggests that we attempt to find it in the following way. Let D_0 be the dominion of R in S; in general, let D_{i+1} be the dominion of R in D_i; at all stages the maximal epic R-subring of R in S must lie in D_i, so we hope that at some stage $D_i = D_{i+1}$, which implies that $R \to D_i$ is an epimorphism and D_i is the maximal epic R-subring of R in S. This is what we shall show and it follows that the maximal epic R-subring of a ring in a semiprimary ring is always semiprimary, since we know that each D_i is semiprimary.

Theorem 7.23 Let $\phi:R \to S$ be a ring homomorphism from R to a semiprimary ring S; then the maximal epic R-subring in S is semiprimary.

Proof: As above, we let D_0 be the dominion of R in S, and in general we let D_{i+1} be the dominion of R in D_i. By theorem 7.19, each D_i is semiprimary. At each stage the number of elements in a maximal set of orthogonal idempotents is finite and decreases as the subscript increases; therefore it is eventually constant at some integer s. Once this stage has been reached, the number of non-isomorphic principal indecomposable projective modules can only increase as the subscript increases, and it is bounded by s; therefore it is eventually constant at some integer t; we assume that we have reached this point for the subring D_m.

Let N_k be the radical of D_k; then for $k > M$, we shall show that $N_k = N_m \cap D_k$. Certainly, $N_m \cap D_k \subseteq N_k$. For the converse, we note that by construction a maximal set of orthogonal idempotents in D_k is also a maximal orthogonal set of idempotents in D_m; this continues to hold for $D_k/N_m \cap D_k$ inside the semisimple artinian ring D_m/N_m. Further the number of non-isomorphic principal indecomposable projectives over $D_k/N_m \cap D_k$ is equal to the number of simple D_m/N_m modules; consequently, if $\{e_i:I = 1 \text{ to } s\}$ is a maximal set of orthogonal idempotents in $D_k/N_m \cap D_k$, $e_iD_k/N_m \cap D_k \cong e_jD_k/N_m \cap D_k$ if and only if $e_iD_m/N_m \cong e_jD_m/N$. If $D_m/N_m \cong \mathop{\times}_i M_{v_i}(E_i)$, for skew fields E_i, it is now clear that

$D_k/N_m \cap D_k \cong \times_i M_{v_i}(F_i)$ where F_i is a semiprimary subring of the skew field E_i and consequently a skew field itself; it follows that $N_k = D_k \cap N_m$.

Suppose that $N_m^{u-1} = 0$. $D_{m+u}/N_{m+u} \subseteq D_{m+1}/N_{m+1}$ D_m/N_m. Since D_{m+1} is the dominion of D_{m+u} in D_m, D_{m+1}/N_{m+1} is contained in the dominion of the semisimple artinian ring D_{m+u}/N_{m+u} in D_m/N_m; so, $D_{m+u}/N_{m+u} = D_{m+1}/N_{m+1}$. In general, assume that $D_{m+u}/N_m^j \cap D_{m+u} = D_{m+j}/N_m^j \cap D_{m+j}$; then $D_{m+u}/N_m^{j+1} \cap D_{m+u} \subseteq D_{m+j}/N_m^{j+1} \cap D_{m+j}$ and they differ only in the socle of the rings which implies that as right modules over $D_{m+u}/N_m^{j+1} \cap D_{m+u}$, $D_{m+j}/N_m^{j+1} \cap D_{m+j} \cong D_{m+u}/N^{j+1} \cap D_{m+u} \oplus M$, where M is a semisimple module; this implies that the dominion of $D_{m+u}/N_m^{j+1} \cap D_{m+n}$ is itself and must equal $D_{m+j+1}/N_m^{j+1} \cap D_{m+j+1}$. For, if $A \subseteq B$ and $B \cong A \oplus M$ as right A module, then $B \otimes_A B \cong A \oplus M \otimes_A B \cong A \otimes_A B \oplus M \otimes_A B$, and $m \otimes 1 \neq 1 \otimes b$ for any m in M and b in B. By induction, we find that $D_{m+u} = D_{m+u-1}$ which proves that D_{m+u-1} is the maximal epic R-subring in S. Therefore, the maximal epic R-subring is semiprimary.

An odd consequence of this result is that a ring has a homomorphism to a simple artinian ring if and only if it has an epimorphism to a simple artinian ring: let $\phi: R \to S$ be some homomorphism to a simple art artinian ring, and let E be the maximal epic R-subring in S; then E is semiprimary and so it maps surjectively to some simple artinian ring, S'; the composite map from R to S' is an epimorphism.

The simple artinian spectrum of a k-algebra

There has been much talk about a theory of non-commutative algebraic geometry. It is not my intention here to add to this, but rather to point out that our preceding theory does give us a functor from rings to topological spaces which is a simple summary of the information on possible homomorphisms from the ring to simple artinian rings. It would be possible to equip this space with a sheaf of rings, and to represent modules over the ring as a sheaf of modules over this sheaf of rings; however, in the absence of any obvious use for this machinery, I shall leave it to future mathematicians of greater insight. Finally, the theory here is stated only for k-algebras over a field k; it is possible to present the theory for arbitrary rings, but the care required in order to avoid simple pitfalls makes the discussion cumbersome.

We begin by considering a particular pre-ordered abelian group functorially associated to a ring R. We define $M(R)$ to be the abelian group generated by isomorphism classes of f.p. modules over R with relations $[A \oplus B] = [A] + [B]$. We define an ordering on $M(R)$ by specifying a positive cone. $[A] > 0$ for all f.p. modules A, and if $A \to B \to C \to 0$ is an exact sequence, we define $[B] - [C] > 0$, and $[A] + [C] - [B] > 0$. If $\phi: R \to S$ is a homomorphism of rings, $[A] \to [A \otimes_R S]$ defines an order-preserving homomorphism from $M(R)$ to $M(S)$, since $\otimes_R S$ is a right exact functor. $M(R)$ has an order unit since there exists n such that $[{}^nR] > [A]$ for any module A. It is now clear that our Sylvester module rank functions on a k-algebra R are simply order-preserving homomorphisms from $M(R)$ to $\frac{1}{m}\mathbb{Z}$ for varying integers m such that $[R]$ goes to 1. Given n such Sylvester module rank functions, ρ_i, we form a family of Sylvester module rank functions $\{\Sigma q_i\rho_i : \Sigma q_i = 1,\ q_i > 0,\ q_i \in \mathbb{Q}\}$. This arises from the ring theory in a natural way, for if the homomorphism $\phi_i: R \to S_i$ gives rise to the Sylvester rank function ρ_i, we have a homomorphism from R to $\times S_i$, and the various homomorphisms from this semisimple artinian ring to simple artinian rings give rise to the rank functions $\Sigma q_i\rho_i$. So, our space of equivalence classes of homomorphisms from R to simple artinian rings has the structure of a $\mathbb{Q}$-convex subset of the space of all order-preserving homomorphisms from $M(R)$ to the reals, and can be given the subspace topology (once the space of all order-preserving maps from $M(R)$ to the reals has been given a suitable topology). A particularly important set of points in this space are the <u>extremal points</u> which are those Sylvester module rank functions that do not lie in the linear span of some other set of Sylvester module rank functions. We should like to be able to show that every Sylvester rank function arises in a unique way as an element in the linear span of extremal rank functions. In order to prove this it is convenient to look at homomorphisms to simple artinian rings in a slightly different way.

If we have a homomorphism from a ring R to a simple artinian ring $M_n(D)$ where D is a skew field, we may regard the simple left $M_n(D)$ module as an R,D bimodule M such that $[M:D] = n$; conversely, given such a bimodule, we automatically have a homomorphism from R to $M_n(D)$. The Sylvester module rank function associated to such a bimodule is given by $\rho(A) = [A \otimes_D M : D]/[M:D]$. We note that the endomorphism ring of M as an R,D

bimodule is the centraliser of R in $M_n(D)$. A decomposition of M as an R,D bimodule, $M \cong M_1 \oplus M_2$ where the rank function associated to M_i is ρ_i allows us to write $\rho = ([M_1:D]\rho_1 + [M_2:D]\rho_2)/[M:D]$. If this happens then either ρ is not extremal or else it is extremal and the image of ρ is a distinct subgroup of $1/[M:D]\rho$. We shall show a sort of converse to this statement.

Theorem 7.24 Any Sylvester rank function lies in the linear span of some set of extremal rank functions.

Proof: Let R be a ring and let ρ be a Sylvester rank function taking values in $\frac{1}{n}\mathbb{Z}$; so ρ arises from a homomorphism from R to $M_n(D)$ for some skew field D; we prove our theorem by induction on n; it is clear for $n = 1$, since all such arise from homomorphisms to skew fields which are extremal points. Let M be the simple left $M_n(D)$ module considered as an R,D bimodule.

Under the assumption that ρ is not extremal, there exist simple artinian rings S_1, S_2 and S and homomorphisms $R \to S_1 \times S_2 \to S$ such that the composite map induces the rank function ρ whilst $R \to S_i$ induces the rank function ρ_i different to ρ. By theorem 7.3, there exists a simple artinian ring $M_n(D) \underset{R}{\circ} S \cong M_{np}(E)$, and from this we form the simple artinian ring $M_{np}(D) \underset{M_i(D)}{\circ} M_{np}(E) \cong M_{np}(F)$; let $\underline{M}$ be the simple left $M_{np}(F)$ module regarded as an R,F bimodule. Since R lies in $M_{np}(D)$, $\underline{M} \cong (M \otimes_D F)^R$ as R,F bimodule; on the other hand, R lies in $S_1 \times S_2$ and so the centraliser of R in $M_{np}(F)$ contains an idempotent e such that the rank function associated to $e\underline{M}$ is the rank function ρ_1; consequently, $e\underline{M}$ is not isomorphic to a direct sum of copies of $M \underset{D}{\otimes} F$; since $\underline{M}$ is a module of finite length, the Remak, Krull, Schmidt theorem implies that $M \otimes_D F \cong M_1 \oplus M_2$ for some R,F bimodules M_1 and M_2. Therefore, ρ lies in the linear span of the rank functions associated to the bimodules M_i; however, $[M_i:F] < n$, and so the theorems follows by induction on n.

By applying the argument used in this theorem in the light of the information it gives us we can strengthen our conclusion yet further.

Theorem 7.25 Any Sylvester rank function has a unique expression as the weighted sum of extremal points; further, if $\rho = 1/m\Sigma m_i\rho_i$ where $\Sigma m_i = m$, h.c.f.$\{m_i\} = 1$, and ρ_i is an extremal rank function mapping onto $1/s_i\mathbb{Z}$, then ρ maps onto $1/\sum_i m_i s_i$.

Proof: Let R be the ring on which we have the Sylvester rank function ρ. Further, let $\rho = 1/m\Sigma m_i\rho_i$ be some expression of ρ as a weighted sum of extremal points ρ_i where ρ_i maps onto $1/s_i\mathbb{Z}$ so that to each rank function ρ_i, we have a homomorphism from R to a simple artinian ring $S_i \cong M_{s_i}(D_i)$ where D_i is a skew field; let M_i be the simple left $M_{s_i}(D_i)$ module regarded as an R,D_i bimodule. Let $R \to M_n(D)$, where D is a skew field, be a homomorphism inducing ρ. Let M be the simple left $M_n(D)$ regarded as R,D bimodule. Let $N = \sum_i s_i m_i$. We also have a homomorphism $R \to \times_i M_{s_i}(D_i) \to \times_i M_{s_i m_i}(D_i) \to M_N(E)$ for some skew field that induces the rank function ρ on R; in order to do this, it must assign the rank $1/n$ to a simple module over the ring $\times_i M_{s_i m_i}(D_i)$, and we we have specified homomorphisms from D_i to E. We form the simple artinian ring $M_N(E) \underset{R}{\circ} M_n(D)$ which is isomorphic to $M_{N_p}(E')$ for some skew field E'; we form the simple artinian ring $M_{N_p}(D) \underset{M_n(D)}{\circ} M_{N_p}(E') \underset{M_N(E)}{\circ} M_{N_p}(E) \cong M_{N_p}(F)$. Let $\underline{M}$ be the simple left $M_{N_p}(F)$ module considered as an R,F bimodule. Since R lies in $\times_i M_{s_i}(D_i)$, its centraliser contains a copy of $M_p(\times M_{m_i}(k))$ where k is the centre of F, and therefore, $\underline{M} \cong \oplus (M_i \otimes_{D_i} F)^{m_i p}$ as R,F bimodule. Since the rank function induced by the bimodule $M_i \otimes_{D_i} F$ is extremal and maps onto $1/s_i\mathbb{Z}$ where $s_i = [M_i : D_i]$, $M_i \otimes_D F$ is indecomposable; however, R also lies in $M_n(D)$, so, on setting $q = Np/n$, we see that $\underline{M} \cong (M\otimes_D F)^q$ as R,F bimodule. By the Remak, Krull, Schmidt theorem, we see that q divides p, and $M\otimes_D F \cong \oplus (M_i \otimes_{D_i} F)^{m_i p/q}$; therefore, N divides n. If ρ maps onto $1/t\mathbb{Z}$ there is a homomorphism to a simple artinian ring $M_t(D')$ where D' is a skew field that induces ρ; we have just shown N must divide t, but it is clear that ρ takes values in $1/N\,\mathbb{Z}$ so that the last statement of the theorem follows. The first statement of the theorem is clear by now, since for any skew field G containing F, $\underline{M}\otimes_F G \cong \oplus (M_i \otimes_{D_i} G)^{m_i p/q}$ is the unique representation of $\underline{M}\otimes_F G$ as a direct sum of indecomposable bimodules, which shows that any representation of ρ as a weighted sum of other rank functions is refinable to this representation in terms of extremal points.

There is one further point we should mention in this discussion of R,D bimodules; the last result shows that the extremal points correspond to bimodules that remain indecomposable under 'extension of skew field'; we should like to be able to characterise those extremal points that arise from epimorphisms to simple artinian rings in some similar way. Before we can do this, we need to link the notion of epimorphisms on rings to the associated functor on categories of modules over the rings. The next result is due to Silver (67). A functor from a category <u>C</u> to a category <u>D</u> is said to be <u>full</u> if it is an embedding, and if M,N are objects in the image of the functor, and α is a map between them, α is in the image of the functor.

<u>Theorem 7.26</u> Let $\phi: R \to S$ be a homomorphism of rings; it is an epimorphism if and only if the forgetful functor ϕ:mod S $\to$ mod R is a full functor.

Proof: If $\phi: R \to S$ is an epimorphism, then for any S module M, $M\otimes_R S \cong M\otimes_S (S\otimes_R S) \cong M\otimes_S S \cong M$: so for any pair of S modules M,N an R linear map from M to N must be S linear as we see by considering the map it induces from $M\otimes_R S$ to $N\otimes_R S$.

Conversely, if $\bar{\phi}$:mod S $\to$ mod R is a full functor, $\mathrm{Hom}_R(S, S\otimes_R S) = \mathrm{Hom}_S(S, S\otimes_R S)$; we have an R linear map from S to $S\otimes_R S$ given by $\alpha(s) = s\otimes 1$; so this defines an S linear map, and $s\otimes 1 = \alpha(s) = \alpha(1)s = 1\otimes s$, which implies that $S\otimes_R S \cong S$, which shows that $R \to S$ is an epimorphism.

The following consequence is noted by Ringel (79).

<u>Theorem 7.27</u> Let R be a ring, and let M be a module over R such that $\mathrm{End}_R(M)$ is a skew field D, and $[M:D] = n$; then the corresponding map from R to $M_n(D)$ is an epimorphism, and all epimorphisms arise in this way. M is the simple $M_n(D)$ module.

Proof: Let $R \to S$ be an epimorphism from R to a simple artinian ring S; then if M is the simple S module, $\mathrm{End}_R(M) = \mathrm{End}_S(M) = D$, a skew field, and $[M:D] = n$ where $S = M_n(D)$.

Conversely, if M is an R module such that $\mathrm{End}_R(M) = D$, a skew field, and $[M:D] = n$, we have a map from R to $M_n(D)$ and given any $M_n(D)$ module, its structure as an R module is a direct sum of copies of

M; so it is clear that all R module maps between S modules are S module maps, which shows that the map from R to $M_n(D)$ is an epimorphism by 8.4.

So this shows that epimorphisms ought to correspond to simple bimodules. Our next theorem proves that this is precisely the case.

Theorem 7.28 Let ρ be a Sylvester rank function on the ring R taking values in $\frac{1}{n}\mathbb{Z}$; then ρ is induced by an epimorphism to a simple artinian ring if and only if there is a skew field D and a simple R,D bimodule M, $[M:D] = n$, that induces this rank function on R.

Proof: If $R \to M_m(D)$ is an epimorphism where D is a skew field then the simple left $M_m(D)$ module M considered as an R,D bimodule is simple.

Conversely, if M is a simple R,D bimodule for some skew field, $[M:D] = m$, let R' be the dominion of R in $M_m(D)$; if N is the radical of R', NM is a distinct R,D sub-module of M; therefore, $N = 0$, and R' is a direct sum of simple artinian rings, since it is a semiprimary ring (theorem 7.19); if e is a central idempotent in R', $e \neq 1$, then eM is a distinct sub-bimodule of M and so $e = 0$. Hence, R' must be simple artinian, and the homomorphism from R to R' is an epimorphism.

The results of this section show that the space of all Sylvester rank functions on a ring form a sort of infinite dimensional $\mathbb{Q}$-simplex since every point in it may be expressed in a unique way as a weighted sum of extremal points; this suggests that some detailed study should be made of the homomorphisms from a ring to simple artinian rings that induce extremal rank functions on it; at present, little has been done in this direction.

In addition to the simplex structure on the space of all Sylvester rank functions, there is also a topology reflecting the notion of specialisation; we define a rank function ρ_1 to be a specialisation of the Sylvester rank function ρ_2 if $\rho_2(M) < \rho_1(M)$ for all f.p. modules M; it is easily checked that this generalises the standard notion of specialisation. The closed sets in our topology on the space of Sylvester rank functions are those that are closed under specialisation. We may also reformulate a version of the support relation discussed by Bergman (76) in terms of the simplex structure on the space of Sylvester rank functions and the specialisation topology. We say that the extremal Sylvester rank function ρ supports the extremal Sylvester rank functions $\{\rho_i : i = 1 \text{ to } n\}$ if it specialises to some point on the interior of the face spanned by $\{\rho_i\}$. Again at present we have little information on this structure.

PART II

Skew Subfields of Simple Artinian Coproducts

8 THE CENTRE OF THE SIMPLE ARTINIAN COPRODUCT

In this chapter, we shall begin a fairly specific study of the simple artinian coproduct with amalgamation; we shall attempt to find the centre of such simple artinian coproducts in terms of the factors and the amalgamated simple artinian subring. Unfortunately, our results are at present incomplete, so we shall begin by stating the conjecture, and then we shall summarise the cases for which it is known to be true, before we prove them.

These results will be used in the next chapter to study the f.d. division subalgebras of a skew field coproduct.

<u>Conjecture 8.1</u> The centre of $S_1 \underset{S}{\circ} S_2$ lies in S, except possibly when both S_1 and S_2 are of rank 2 over S; in this case, the centre lies in S, or is the function field of a curve of genus 0 over its intersection with S.

The <u>function field of a curve of genus 0</u> over a field K is a field F such that $L \otimes_K F \cong L(t)$, where L is a suitable f.d. extension of K, and t is a transcendental. We shall present examples to show that all curves of genus 0 arise as centres of skew field coproducts at the end of this chapter.

This conjecture is known to be true when S is a common central subfield of S_1 and S_2. It is also true for skew field coproducts amalgamating an arbitrary skew subfield, except possibly when there are only two factors both of which are of dimension 2 over the amalgamated skew subfield. It may also be proven if one of the factors is simple artinian.

The result that we use to prove our conjecture when we can is due to Cohn (84); it summarises the connection between the centre of a fir, and the centre of its universal skew field.

Theorem 8.2 Let T be a non-Ore fir with universal skew field of fractions U; then the centre of U is the centre of T.

The proof is too long to include here; the interested reader should look at Cohn (84). Our first special case of 8.1 is a direct application of this.

Theorem 8.3 The centre of $E_1 \circ_E M_m(E_2)$ lies in E, provided that $[E_1:E]$ or $[M_m(E_2):E]$ is greater than 2.

Proof: $E_1 \circ_E M_m(E_2)$ is the universal simple artinian ring of fractions of $E_1 \sqcup_E M_m(E_2) \cong M_m(R)$, where R is a fir, since, by Bergman's coproduct theorems (see chapter 3), and the Morita equivalence of $M_m(R)$ and R, every submodule of a free R module is free. Further, the assumptions on dimension imply that R cannot be Ore, and so, by 10.2, the centre of the universal skew field of fractions of R is the centre of R. So, the centre of the universal simple artinian ring of fractions of $M_m(R)$ is the centre of $M_m(R)$, which lies in R, and, in particular, consists of units of R.

R is the ring of endomorphisms of the $M_m(R)$ module $S \otimes_{M_m(E_2)} M_m(R)$, where S is the unique simple $M_m(E_2)$ module, and the units of R induce the automorphisms of this module. By 2.18, the group of automorphisms of this module are induced by the automorphisms of S as $M_m(E_2)$ module; so the units of R lie in the copy of E_2 inside it; therefore, the centre of R lies in the copy of E_2 inside it, and the centre of $M_m(R) \cong E_1 \sqcup_E M_m(E_2)$ must also lie in E_2. However, if e is an element of $M_m(E_2) - E$, and f is an element of $E_1 - E$, $ef \neq fe$, so that e cannot be central. Therefore, the centre of $M_m(R)$ which lies in E_2 actually lies in E. But this is the centre of its universal simple artinian ring of fractions, as we saw at the end of the last paragraph.

Since this is all that is used in the following chapter, and the remaining proofs of results towards conjecture 8.1 are complicated, the reader may wish to bypass them for the time being; in this case, he may wish to look only at the examples which show that all function fields of curves of genus 0 arise as the centre of a suitable simple artinian coproduct.

We wish to find the centre of the simple artinian coproduct of two simple artinian rings over a common central subfield. The next two theorems deal first with the non-Ore, and then with the Ore case. Of the two, the non-Ore case is more complex.

<u>Theorem 8.4</u> The centre of $M_{m_1}(E_1) \circ_k M_{m_2}(E_2)$ is equal to k, provided that one of $[M_{m_1}(E_1):k]$ or $[M_{m_2}(E_2):k]$ is greater than 2.

Proof: The idea of the proof is to reduce to the previous theorem by replacing $M_{m_1}(E_1)$ by a skew field closely related to it; more precisely, the skew field will be a twisted form of $M_{m_1}(E_1)$. A simple artinian ring S is said to be a <u>twisted form</u> of a simple artinian ring S', if we have a central subfield k of each of them, and a f.d. field extension L of k such that $S \otimes_k L \cong S' \otimes_k L$. For a detailed discussion of twisted forms of algebraic structures, the reader should consult Waterhouse (79).

A twisted form of $M_{m_1}(E_1)$ that is a skew field need not exist; however, the case where one does exist is an important special case:

<u>Special case</u>: there exists a central division algebra D such that $[D:k] = m_1$, $D \otimes_k E_1$ is a skew field, and D has a Galois splitting field L such that $E_i \otimes_k L$ is simple artinian for $i = 1,2$.

In this case, our idea is to show that $((D \otimes_k E_1) \circ_k M_{m_2}(E_2)) \otimes_k L \cong (M_{m_1}(E_1) \circ_k M_{m_2}(E_2)) \otimes_k L$. From theorem 10.3, we know that $(D \otimes_k E_1) \circ_k M_{m_2}(E_2)$ has centre k; if C is the centre of $M_{m_1}(E_1) \circ_k M_{m_2}(E_2)$, it follows from the isomorphism that $C \otimes_k L \cong L$, so $C \cong k$.

Consider the ring $D \otimes_k E_1 \circ_k M_{m_2}(E_2)$; then by theorem 8.3, it has centre k. Therefore, $S \cong (D \otimes_k E_1 \circ_k M_{m_2}(E_2)) \otimes_k L$ is simple artinian and it is a universal localisation of $(D\otimes_k E_1 \underset{k}{\sqcup} M_{m_2}(E_2))\otimes_k L \cong (D\otimes_k E_1 \otimes_k L) \underset{L}{\sqcup} M_{m_2}(E_2\otimes L)$ which is isomorphic to $M_{m_1}(E_1\otimes_k L) \underset{L}{\sqcup} M_{m_2}(E_2\otimes_k L)$ an hereditary ring with unique rank function, since each of the factors is simple artinian. On the other hand, this ring is visibly isomorphic to $(M_{m_1}(E_1) \underset{k}{\sqcup} M_{m_2}(E_2))\otimes_k L$, which has the universal localisation $(M_{m_1}(E_1) \underset{k}{\circ} M_{m_2}(E_2))\otimes_k L$ which is semisimple artinian and hence weakly finite, and so must be a subring of S. But S must be epic over it, and so it is actually the whole of S. Since S has centre L, the centre of $M_{m_1}(E_1) \underset{k}{\circ} M_{m_2}(E_2)$ is k.

<u>General case</u>: In order to deal with the situation where there is no f.d. division algebra D of the sort required for the application of the special case, we extend the centre k until there is.

Let K be the commutative field $k(x_1,\ldots,x_m)$, where $\{x_i\}$ is a set of commuting indeterminates. There is an automorphism ϕ of this field of order m_1, given by the cyclic permutation of the indices. We form the skew field $D = K(Y:\phi)$; let C be the centre of D.

It is easy to see that $D\otimes_k E$ is a Noetherian domain for any skew field E, and so, $D\otimes_k E$ has a skew field of fractions.

Let E_i' be the skew field of fractions of $E_i\otimes_k C$ and let F' be the skew field of fractions of $F\otimes_k C$, where $M_m(F) \cong M_{m_1}(E_1) \underset{k}{\circ} M_{m_2}(E_2)$. Then we have the equation $M_m(F') \cong M_{m_1}(E_1') \underset{C}{\circ} M_{m_2}(E_2')$. If L is central in $M_{m_1}(E_1) \underset{k}{\circ} M_{m_2}(E_2)$, then $L\otimes_k C$ is central in $M_{m_1}(E_1') \underset{C}{\circ} M_{m_2}(E_2')$. So, once we know that the centre of $M_{m_1}(E_1') \underset{k}{\circ} M_{m_2}(E_2')$ is C, we shall know that the centre of $M_{m_1}(E_1) \underset{k}{\circ} M_{m_2}(E_2)$ is k.

However, $[D:C] = m_1^2$, $D\otimes_k E_1'$ is the skew field of fractions of $D\otimes_k E_1$, and $K\otimes_C E_i$ is actually a skew field for $1 = 1,2$, so, by our first case, the centre of $M_{m_1}(E_1') \underset{C}{\circ} M_{m_2}(E_2')$ is C, which forces the centre of $M_{m_1}(E_1) \underset{k}{\circ} M_{m_2}(E_2)$ to be k, as we wished to show.

We wish to deal with the remaining case of the coproduct of two simple artinian rings S_1 and S_2 such that $[S_i:k] = 2$; clearly, both are commutative fields.

<u>Theorem 8.5</u> Let E_1 and E_2 be a pair of quadratic extensions of k. Then the centre of $E_1 \underset{k}{\circ} E_2$ is purely transcendental of transcendence degree 1 over k.

Proof: This was also shown by Cohn by a different method.

By Bergman (74') or a simple calculation, the centre of $E_1 \underset{k}{\sqcup} E_2$ has the form $k[t]$, and $E_1 \underset{k}{\sqcup} E_2$ is a free module of rank 4 over $k[t]$. $E_1 \underset{k}{\circ} E_2$ must be the central localisation of $E_1 \underset{k}{\sqcup} E_2$, and must be of rank 4 over the central subfield $k(t)$. Since it is not commutative, this must be the whole centre.

We finish off this chapter by showing that all function fields of curves of genus 0 over k occur as the centre of a suitable simple artinian coproduct with amalgamation. We shall need the next lemma in the proof of this.

<u>Lemma 8.6</u> $M_2(L) \underset{L\times L}{\sqcup} M_2(L) \cong M_2(L[t,t^{-1}])$ for a commuting indeterminate t.

<u>Proof</u>: All embeddings of $L \times L$ in $M_2(L)$, are conjugate to the embedding along the diagonal; so we may assume that the first factor has matrix units e_{11}, e_{12}, e_{21}, and e_{22}, whilst the second factor has matrix units e_{11}, f_{12}, f_{21}, and e_{22}.

Since f_{12} lies in $e_{11}(M_2(L) \underset{L\times L}{\sqcup} M_2(L))e_{22}$, it takes the form $\begin{pmatrix} 0 & t \\ 0 & 0 \end{pmatrix}$, when it is written as a matrix over the centraliser of the first factor, and similarly, f_{21} has the form $\begin{pmatrix} 0 & 0 \\ s & 0 \end{pmatrix}$, from which it is clear that $s = t^{-1}$. Since the centraliser of the first factor is generated by the entries of these matrices, the centraliser is isomorphic to $L[t,t^{-1}]$ where t is an indeterminate, since the dimensions of the coproduct we are considering is infinite.

Suppose that F is the function field of a curve of genus 0 over k; that is, $F \otimes_k K \cong K(t)$ for some Galois extension K of k with Galois group G. Then F is a twisted form of $k(t)$ and so by twisted form theory, it corresponds to a suitable element of $H^1(G, PGL_2(K))$ since $PGL_2(K)$ is the automorphism group of $k(x)$ over K (for example, see Waterhouse (79)). Since $PGL_2(K)$ is also the automorphism group of $M_2(K)$ over K, this element α of $H^1(G, PGL_2(K))$ corresponds to some central quaternion algebra D_F over k; moreover, Roquette (62) shows that F is the universal splitting field of D_F. Let L be a maximal separable commutative subfield of D_F; $[L:k] = 2$.

<u>Theorem 8.7</u> Name everything as in the last paragraph; then $M_2(k) \underset{L}{\circ} D_F \cong M_2(F)$; so all function fields of curves of genus 0 arise as the centre of a suitable simple artinian coproduct with amalgamation.

Proof: $(M_2(k) \underset{L}{\sqcup} D_F) \otimes_k L \cong M_2(L) \underset{L \times L}{\sqcup} M_2(L) \cong M_2(L[t, t^{-1}])$ by the lemma above. So, $(M_2(k) \underset{L}{\circ} D_F) \otimes_k L \cong M_2(L(t))$, since both sides of this equation are obtained by central localisation of the corresponding sides in the first equation.

Let F' be the centre of $M_2(k) \underset{L}{\circ} D_F$; then $F' \otimes_k L \cong L(t)$, so that F' is either purely transcendental or else it is the splitting field of a suitable quaternion algebra $D_{F'}$. However, $M_2(k) \underset{L}{\circ} D_F \cong F' \otimes_k D_F$, so that F' cannot be purely transcendental since it splits D_F. Roquette (62) shows that the universal splitting field of a central quaternion algebra can split no other central simple algebra, so F' must be isomorphic to F, which is what we want.

9 FINITE DIMENSIONAL DIVISION SUBALGEBRAS OF SKEW FIELD COPRODUCTS

We intend in this chapter to make a detailed study of division subalgebras of skew field coproducts; since the method developed may be applied in much greater generality, we shall develop the results in this context and then specialise the results to skew field coproducts. In later sections of this chapter, we shall briefly consider its applications to related rings.

Division subalgebras of universal localisations

For the purposes of this section, we shall assume that the rings R,F,K,k satisfy the following conditions:

Assumptions: R is a left semihereditary k-algebra, where k is a commutative field. ρ is a rank function on the f.g. projectives over R taking values in $\frac{1}{n}\mathbb{Z}$ such that the universal localisation of R at the rank function, R_ρ, is a simple artinian ring $M_n(F)$ where F is a skew field with centre K, where K is a regular extension of k.

On the whole, there is nothing to this beyond naming the rings that we are interested in.

An extension of commutative fields $K \supset k$ is said to be regular if $K\otimes_k L$ is a domain for all commutative field extensions of k.

It is useful to reformulate our problem; we wish to find whether a f.d. division algebra D over k embeds in a simple artinian ring $M_n(F)$ with centre K, a regular extension of k. We can broaden this a little and ask for what numbers does D embed in $M_m(F)$. This has a useful reformulation.

Lemma 9.1 Under the previous assumptions, $D^o\otimes_k F$ is a simple artinian ring. Let S be the simple module for $D^o\otimes_k F$, and let $[S:F] = s$; then D embeds in $M_m(F)$ if and only if s divides m.

Proof: Let C be the centre of D; then $D^o \otimes_k F \cong D^o \otimes_C (C \otimes_k F)$, so any non-zero ideal intersects non-trivially with $C \otimes_k F$. In turn, $C \otimes_k F \cong (C \otimes_k K) \otimes_K F$, so any non-zero ideal of $C \otimes_k F$ intersects non-trivially with $C \otimes_k K$ which is a field by our assumption on K. So, $D^o \otimes_k F$ must be simple artinian.

We are left with the last sentence of the lemma. If D embeds in $M_m(F)$, D^o embeds in $M_m(F^o)$, which is the endomorphism ring of the F module F^m. Consequently, we can give F^m the structure of a left $F \otimes_k D^o$ module, and so, F^m is a direct sum of copies of S. The argument is reversible.

It is important to bear this in mind, as most of the embedding theorems will be stated in terms of the dimension over F of a simple $D^o \otimes_k F$ module, since this actually gives us information about all possible embeddings of D in matrix rings over F. It is also worth remembering that the same information is conveyed by the rank as $M_m(F)$ module of a simple $M_m(F) \otimes_k D^o$ module, since the Morita equivalence of $M_m(F) \otimes_k D^o$ and $F \otimes_k D^o$ shows that the dimension of a simple $F \otimes_k D^o$ module over F is m times the rank of a simple $M_m(F) \otimes_k D^o$ as $M_m(F)$ module.

<u>Theorem 9.2</u> Suppose that our previous assumptions hold, and let D be a f.d. division algebra over k; then the rank of a simple $R_\rho \otimes_k D^o$ module over R_ρ is equal to $\text{h.c.f.}_M\{\rho(M)\}$ as M runs through f.g. left $R \otimes_k D^o$ submodules of free $R \otimes_k D^o$ modules.

Proof: That the rank of a simple $R_\rho \otimes_k D^o$ module as an R_ρ module must divide this h.c.f. is easily seen by considering the $R_\rho \otimes_k D^o$ modules, $R_\rho \otimes_R M$, where M is an $R \otimes_k D^o$ module. We are left with the converse.

First, we see that $R_\rho \otimes_k D^o$ is a universal localisation of $R \otimes_k D^o$ at the full maps with respect to ρ over the ring R. So we can use Cramer's rule. Let e be an idempotent in $R_\rho \otimes_k D^o$ such that $(R_\rho \otimes_k D^o)e$ is isomorphic to the unique simple module S over $R_\rho \otimes_k D^o$. Then, by 4.3, e is stably associated to some map between f.g. projectives induced up from some map between f.g. projectives over $R \otimes_k D^o$, $\alpha: Q \to P$; we may choose this map to be associated to $\begin{matrix} I_t & 0 \\ 0 & e \end{matrix}$. The image of $\begin{matrix} I_t & 0 \\ 0 & e \end{matrix}$ is isomorphic to $(R_\rho \otimes_k D^o)^t \oplus S$, so this is also isomorphic to the image of our map $R_\rho \otimes_R \alpha$

between f.g. projectives over $R_\rho \otimes_k D^o$.

Let $M \subseteq P$ be the image of α over $R \otimes_k D^o$; then the image of this map extended to $R_\rho \otimes_k D^o$ is $(R_\rho \otimes_k D^o)M \subseteq (R_\rho \otimes_k D^o) \otimes_{R_\rho \otimes_t D^o} P$. Since D^o centralises the action of R and R_ρ, we may actually write this equation as $R_\rho M \subseteq R_\rho \otimes_R P$. At this point it is useful to recall that 1.10 shows that if $N \subseteq R^s$ is a f.g. submodule of R^s, the rank of $R_\rho N$ as R_ρ module is equal to the minimal rank of R modules N' such that $N \subseteq N' \subseteq R^s$. In the case we are examining, we have an $R \otimes_k D^o$ module $M \subseteq P$, and we wish to find the rank as R_ρ module of $R_\rho M \subseteq R_\rho \otimes_R P$, since we know that $R_\rho M$ is isomorphic to $(R_\rho \otimes_k D^o)^t \oplus S$, and we shall be able to deduce from this the rank of S as an R_ρ module. We know by 1.10 that the rank of $R_\rho M$ as R_ρ module is the minimal rank of an R module M' such that $M \subseteq M' \subseteq P$; our final step is to show that there is an $R \otimes_k D^o$ module above M having this minimal rank; in fact, we shall see that if M' has the minimal rank, so does $D^o M'$ which is an $R \otimes_k D^o$ module since D^o centralises R.

Let $\{1 = d_0, d_1 \ldots d_m\}$ be a basis of D^o over k; then $D^o M' = \Sigma d_i M'$. $d_i M'$ is isomorphic to M' as R module because the action of D^o commutes with that of R, and $d_i M'$ contains M since M is an $R \otimes_k D^o$ module. If M_1 and M_2 are both submodules of minimal rank q in ρ containing M, consider the exact sequence:

$$0 \to M_1 \cap M_2 \to M_1 \oplus M_2 \to M_1 + M_2 \to 0$$

which is split as a sequence of R modules since R is left semihereditary. So, $2q = \rho(M_1 \cap M_2) + \rho(M_1 + M_2)$; but both terms on the right are at least q, which implies that they are exactly q. So $\rho(M_1 + M_2) = q$, and an easy induction shows that $\rho(D^o M') = \rho(\Sigma d_i M') = q$.

We see that the rank as an R_ρ module of $(R_\rho \otimes_k D^o)^t \oplus S$ is equal to the rank as an R module of the $R \otimes_k D^o$ module, $D^o M'$, which is a f.g. submodule of a free module. Consequently, h.c.f.$_M\{\rho(M)\}$ as M runs through the f.g. $R \otimes_k D^o$ submodules of free modules divides the rank of S as an R_ρ module. The converse has already been noted.

Division subalgebras of simple artinian coproducts

In this section, we apply 9.2 to the problem of finding the f.d. division subalgebras of a simple artinian coproduct amalgamating a simple artinian subring under the assumption that conjecture 8.1 holds; since we have verified this conjecture in chapter 8 for a number of cases including the simple artinian coproduct amalgamating a central subfield, this is not unreasonable.

There is a spurious generality in dealing with the amalgamation over a simple artinian ring, since this may be reduced to amalgamation over a skew field instead; whilst performing this reduction we set up the notation and names for this section.

Let $S \cong M_n(E)$, and $S_i \cong M_{n_i}(E_i)$ for $i = 1,2$, where S and S_i are simple artinian rings and S lies in S_i, and E_i is a skew field. Since S embeds in S_i, n divides n_i, so let $m_i = n_i/n$; then on taking the centraliser of $M_n(k)$ in $S_1 \sqcup_S S_2$ we see that $S_1 \sqcup_S S_2 \cong M_n(M_{m_1}(E_1) \sqcup_E M_{m_2}(E_2))$, so we need only consider the ring

$M_{m_1}(E_1) \circ_E M_{m_2}(E_2)$, since $S_1 \circ_S S_2$ is the n by n matrix ring over it. We shall call the rank function on $M_{m_i}(E_i)$ modules ρ_i and the unique rank function on R modules where $R = M_{m_1}(E_1) \sqcup_E M_{m_2}(E_2)$ will be called ρ. Let $m = \text{l.c.m.}\{m_1, m_2\}$; then ρ takes values in $\frac{1}{m}\mathbb{Z}$, and the universal localisation of R at ρ, R_ρ, was shown in 5.6 to be a simple artinian ring, $M_m(F)$, where F is a skew field; we use ρ' for the rank function on $M_m(F)$ modules.

Let C be the intersection of the centre of $M_m(F)$ with E; we recall that our conjecture states that either this is the centre of our coproduct, or else, the centre is a function field of a curve of genus 0 over C. At any rate, we are assuming that it is a regular extension of C.

<u>Theorem 9.3</u> We use the preceding notation; further, we assume that the simple artinian coproduct considered satisfies conjecture 8.1. Let D be a f.d. division algebra over C; then the rank of a simple $D^o \otimes_C M_m(F)$ module as $M_m(F)$ module is $\text{h.c.f.}_{i,j}\{\rho_i(A_{ij})\}$ as A_{ij} runs through f.g. $D^o \otimes_C M_{m_i}(E_i)$ modules.

Proof: We have $R = M_{m_1}(E_1) \sqcup_E M_{m_2}(E_2)$ and $R_\rho \cong M_{m_1}(E_1) \circ_E M_{m_2}(E_2) \cong M_m(F)$. Our conditions fulfil those required for theorems 9.2 to apply. Therefore, the rank of a simple $M_m(F) \otimes_C D^o$ module is $\text{h.c.f.}_M\{\rho(M)\}$ as M runs through f.g. $R\otimes_k D^o$ submodules of free $R\otimes_k D^o$ modules. In order to change this to the form of our theorem, we have to deal with a special case.

Special case: Assume that E has centre C.

Then $R\otimes_C D^o \cong (M_{m_1}(E_1) \underset{E}{\sqcup} M_{m_2}(E_2))\otimes_C D^o \cong (M_{m_1}(E_1)\otimes_C D^o) \underset{E\otimes_C D^o}{\sqcup} (M_{m_2}(E_2)\otimes_C D^o)$. Since E has centre C, $E\otimes_C D^o$ is simple artinian, which allows us to use Bergman's coproduct theorems. By 2.2, any $R\otimes_C D^o$ submodule of a free module has the form $M_1\otimes_{(M_{m_1}(E_1)\otimes_C D^o)}(R\otimes_C D^o) \oplus M_2\otimes_{(M_{m_2}(E_2)\otimes_C D^o)}(R\otimes_C D^o)$, where M_i is a submodule of a free $M_{m_i}(E_i)\otimes_C D^o$ module. We see that $\rho(M) = \rho_1(M_1) + \rho_2(M_2)$, and so, since the rank of a simple $M_m(F)\otimes_C D^o$ module as $M_m(F)$ module equals $\text{h.c.f.}_M\{\rho(M)\}$ as M runs through f.g. submodules of free $R\otimes_C D^o$, it is also equal to $\text{h.c.f.}_{i,j}\{\rho_i(B_{ij})\}$ as B_{ij} runs over f.g. submodules of free $M_{m_i}(E_i)\otimes_C D^o$ modules.

Given any f.g. $M_{m_i}(E_i)\otimes_C D^o$ module, A_{ij}, it has a presentation:

$$0 \to B_{ij} \to s(M_{m_i}(E_i)\otimes_C D^o) \to A_{ij} \to 0$$

and so, it is clear that $\text{h.c.f.}_{ij}\{\rho_i(A_{ij})\}$ as A_{ij} runs through every f.g. module over $M_{m_i}(E_i)\otimes_C D^o$ is equal to $\text{h.c.f.}\{\rho_i(B_{ij})\}$ as B_{ij} runs through f.g. submodules of free modules.

Before passing to the general case, we note the following corollary which is of importance for a reduction to the case we have just dealt with.

Corollary 9.4 Let S be a simple artinian C algebra, and let D be a f.d. division algebra over C; then the rank of a simple $(S\underset{C}{\circ}C(x))\otimes_C D^o$ module as $S\underset{C}{\circ}C(x)$ module is $\text{h.c.f.}_j\{\rho_S(A_j)\}$, where ρ_S is the rank function on S modules, and A_j runs through the f.g. $S\otimes_C D^o$ modules. In particular, if the centre of S is a regular extension of C, the rank of a simple $(S\underset{C}{\circ}C(x))\otimes_C D^o$ module over $S\underset{C}{\circ}C(x)$ equals the rank of a simple $S\otimes_C D^o$ module as S module.

Proof: We saw in theorem 8.3 that the centre of $S \circ_C C(x)$ lies in C, and so, must be C; therefore, we can apply the first case to prove our corollary. The last sentence follows since $S \otimes_C D^o$ is simple artinian in this instance.

General case

We return to the general case. We assumed that the centre of $M_{m_1}(E_1) \circ_E M_{m_2}(E_2)$ is either C or else the function field of a curve of genus 0 over C. In either case, it is a regular extension of C, and so, by the last part of 9.4, the rank of a simple $(M_{m_1}(E_1) \circ_E M_{m_2}(E_2)) \otimes_C D^o$ module as $M_{m_1}(E_1) \circ_E M_{m_2}(E_2)$ module is the rank of a simple $(M_{m_1}(E_1) \circ_E M_{m_2}(E_2) \circ_C C(x)) \otimes_C D^o$ module as an $M_{m_1}(E_1) \circ_E M_{m_2}(E_2) \circ_C C(x))$ module.

We shall show that $M_{m_1}(E_1) \circ_E M_{m_2}(E_2) \circ_C C(x)$ is isomorphic to the ring $(M_{m_1}(E_1) \circ_C C(x)) \underset{E \circ_C C(x)}{\circ} (M_{m_2}(E_2) \circ_C C(x))$.

We note the two factors in this coproduct and also the amalgamated skew field all have centre C bg 9.3, so that we may apply 9.4 to each of the factors, and then our first case to the whole coproduct in order to find the rank of a simple $(M_m(F) \circ_C C(x)) \otimes_C D^o$ as $M_m(F) \circ_C C(x)$ module, which is the rank of a simple $M_m(F) \otimes_C D^o$ module as $M_m(F)$ module, the number we wish to find.

$M_{m_1}(E_1) \circ_E M_{m_2}(E_2) \circ_C C(x)$ is the universal localisation of the hereditary ring $T = M_{m_1}(E_1) \sqcup_E M_{m_2}(E_2) \sqcup_C C[x]$ at the unique rank function on T.

Consider the subring $M_{m_i}(E_i) \sqcup_C C[x] \cong M_{m_i}(E_i) \sqcup_E (E \sqcup_C C[x])$, which has the universal localisation $M_{m_i}(E_i) \circ_E (E \circ_C C(x))$. Since $M_{m_i}(E_i) \sqcup_C C[x]$ has a unique universal localisation that is simple artinian, this shows the isomorphism $M_{m_i}(E_i) \circ_C C(x) \cong M_{m_i}(E_i) \circ_E (E \circ_C C(x))$.

Let Σ_i for $i = 1,2$ be the collection of full maps between f.g. projectives with respect to the unique rank function on $M_{m_i}(E_i) \sqcup_C C[x]$, and let Σ be the union of these maps induced up to T. We consider the ring T_Σ and the ring T', where T' is the ring $M_{m_i}(E_1) \otimes_C C(x)) \sqcup_{E \otimes_C C(x)} (M_{m_2}(E_2) \otimes_C C(x))$. We show that T_Σ is isomorphic to T'.

There is a homomorphism from T_Σ to T', sending $M_{m_i}(E_i)$ to $M_{m_i}(E_i)$ by the identity map and x to x. Under this homomorphism, all elements of Σ are inverted, so that it extends to a homomorphism from T_Σ onto T', since its image contains $M_{m_i}(E_i) \otimes_C C(x)$ for $i = 1,2$.

Conversely, T_Σ is generated by a copy of $M_{m_i}(E_1) \otimes_C C(x)$ and a copy of $M_{m_2}(E_2) \otimes_C C(x)$, which have as common subring $E \otimes_C C(x)$; therefore, there is a homomorphism from T' to T_Σ which is clearly inverse to the homomorphism in the last paragraph.

T' has the universal localisation $M_{m_1}(E_1) \otimes_C C(x)) \otimes_{E \otimes_C C(x)} (M_{m_2}(E_2) \otimes_C C(x))$ which must be the universal localisation of T that is simple artinian; this is $M_{m_1}(E_1) \otimes_E M_{m_2}(E_2) \otimes_C C(x)$, which proves the isomorphism we need.

Since $E \otimes_C C(x)$ has centre C, we may apply the first part of the proof of this theorem to conclude that a simple $(M_{m_1}(E_1) \otimes_C C(x)) \otimes_{E \otimes_C C(x)} (M_{m_2}(E_2) \otimes_C C(x)) \otimes_C D^o$ module has rank as $(M_{m_1}(E_1) \otimes_C C(x)) \otimes_{E \otimes_C C(x)} (M_{m_2}(E_2) \otimes_C C(x))$ module equal to h.c.f. $\{\bar{\rho}_1(A_1), \bar{\rho}_2(A_2)\}$ where A_i is the unique simple $(M_{m_i}(E_i) \otimes_C C(x)) \otimes_C D^o$ module, and $\bar{\rho}_i$ is the rank function on $(M_{m_i}(E_i) \otimes_C C(x))$ modules.

By 9.4, $\rho_i(A_i)$ is equal to h.c.f.$_j\{\rho_i(A_{ij})\}$ as A_{ij} runs over f.g. $M_{m_i}(E_i)\otimes_C D^o$ modules.

Putting this together, we find that the rank of a simple $(M_m(F) \underset{C}{\circ} C(x))\otimes_C D^o$ module as $M_m(F)$ module is h.c.f.$_{i,j}\{\rho_i(A_{ij})\}$ as A_{ij} runs through f.g. $M_{m_i}(E_i)\otimes_C D^o$ modules. We have already remarked that if conjecture 8.1 holds for $M_{m_1}(E_1) \underset{E}{\circ} M_{m_2}(E_2)$, then this number is the same as the rank of a simple $M_m(F)\otimes_C D^o$ module as $M_m(F)$ module by 8.4. This completes the proof.

As we saw in 9.1, this result gives us all the embeddings of D in $M_s(F)$ for varying s. The major part of the rest of this chapter illustrates what exactly this theorem means. Before ending this section, we notice the following simple corollary.

Theorem 8.5 Suppose that the simple artinian coproduct $M_{m_1}(E_1) \underset{E}{\circ} M_{m_2}(E_2)$ satisfies the conditions of theorem 8.3; then the rank of a simple $D^o\otimes_C(M_{m_1}(E_1) \underset{E}{\circ} M_m(E_2))$ module over $M_{m_1}(E_1) \underset{E}{\circ} M_{m_2}(E_2)$ is the rank of a simple $D^o\otimes_C(M_{m_1}(E_1) \underset{C}{\circ} M_{m_2}(E_2))$ module over $M_{m_1}(E_1) \underset{C}{\circ} M_{m_2}(E_2)$.

Proof: Compare the formulas given by 8.3; they are identical.

Finally, we could have stated 8.3 for the simple artinian coproduct of arbitrarily many simple artinian E-rings $M_{m_i}(E_i)$ provided that l.c.m. $\{m_i\}$ exists. The proof does not differ from that given in 8.3.

Division subalgebras of skew field coproducts

Here, we shall discuss more explicitly the consequences of theorem 9.3; we shall organise this by discussing when a division subalgebra of a skew field coproduct must be conjugate to a division subalgebra of one of the factors. This is in analogy to what is known to hold for group coproducts. It usually fails here, but our investigation does have some interesting consequences.

By 9.5, we may as well restrict our attention to skew field coproducts over a central subfield k, since the division subalgebras of skew field coproducts amalgamating other skew subfields may be reduced to this case whenever we can calculate them. By 8.3, the only case that at present eludes us is the skew field coproduct of two skew fields E_1 and E_2 both of which are of dimension 2 on either side over the amalgamated skew subfield E. For much of this section, we shall further restrict our attention to the case where each of the factors of the skew field coproduct over the central subfield k are f.d. over k.

Theorem 9.6 Let $\{E_i: i = 1 \text{ to } n\}$ be f.d. division algebras over k, and let $t = \text{l.c.m.}_i\{[E_i:k]\}$. If D is a f.d. division subalgebra of $\underset{k}{\circ} E_i$, then $[D:k]$ divides t.

Proof: $[D:k]$ divides $[A_{ij}:k]$ as A_{ij} runs through f.g. $D^{\circ}\otimes_k E_i$ modules. $[A_{ij}:k] = [A_{ij}:E_i][E_i:k]$ which divides $[A_{ij}:E_i]t$. If D embeds in $\underset{k}{\circ} E_i$, $\text{h.c.f.}_{i,j}\{[A_{ij}:k]\} = 1$, and so, $[D:k]$ divides t.

The next result has slightly more general hypotheses, and clearly applies to the cases we are considering.

Theorem 9.7 Let $\{E_i: i = 1 \text{ to } n\}$ be skew fields satisfying a polynomial identity, and let D be a f.d. division subalgebra of $\underset{k}{\circ} E_i$; then the polynomial identity degree of D divides $\text{l.c.m.}_i\{\text{p.i. degree } E_i\}$.

Proof: Let $p = \text{l.c.m.}_i\{\text{p.i.degree } E_i\}$. Let A_{ij} be a $D^{\circ}\otimes_k E_i$ module, and let $m_{ij} = [A_{ij}:E_i]$. Then the centraliser of the action of E_i on A_{ij} is isomorphic to $M_{m_{ij}}(E_i^{\circ})$, and D° embeds in it. By Bergman, Small (75), p.i.degree D = p.i.degree D° must divide p.i.degree $M_{m_{ij}}(E_i)$ which is just m_{ij}(p.i.degree E_i). If D embeds in $\underset{k}{\circ} E_i$, $\text{h.c.f.}_{i,j}\{M_{ij}\} = 1$; so that p.i.degree D divides p.

Our next theorem is more by way of example to show that the bounds in 9.6 and 9.7 are in general best possible.

Theorem 9.8 Let D_1 and D_2 be f.d. division algebras over k such that $[D_1:k]$ is co-prime to $[D_2:k]$; then $D_1\otimes_k D_2$ embeds in $D_1 \underset{k}{\circ} D_2$.

Proof: We have a simple $D_i \otimes_k (D_1 \otimes_k D_2)^o$ module of dimension $[D_j:k]$ over D_i for $i \neq j$; by 9.3, there is a simple $(D_1 \underset{k}{\circ} D_2) \otimes_k (D_1 \otimes_k D_2)^o$ module of dimension 1 over $D_1 \underset{k}{\circ} D_2$; that is, $D_1 \otimes_k D_2$ embeds in $D_1 \underset{k}{\circ} D_2$.

Our next result in contrast to the last one gives us a number of cases where a division subalgebra of a skew field coproduct must be conjugate to a division subalgebra of one of the factors.

<u>Theorem 9.9</u> Let E be a f.d. (possibly non-commutative) Galois extension of k; then any f.d. division subalgebra of $E \underset{k}{\circ} k(x)$ is conjugate to a division subalgebra of E.

Proof: Let C be the centre of E; then C is a Galois extension of k with Galois group G, where G is the group of outer automorphisms of E over k.

$D^o \otimes_k C \cong \underset{i}{\times} S_i$, where each S_i is a simple algebra over C, and the group G acts transitively on this set.

$D^o \otimes_k E \cong (D^o \otimes_k C) \otimes_C E \cong (\underset{i}{\times} S_i) \otimes_C E \cong \underset{i}{\times} (S_i \otimes_C E)$, and the group G permutes the simple algebras $S_i \otimes_C E$ transitively.

Therefore, $[A_i:E] = [A_j:k]$ where A_i is a simple $S_i \otimes_C E$ module, and the dimension of a simple $D^o \otimes_k (E \underset{k}{\circ} k(x))$ module must equal the dimension of any simple $D^o \otimes_k E$ module, by 9.3, since they are all the same. So D embeds in $E \underset{k}{\circ} k(x)$ if and only if D embeds in E.

Since the Noether, Skolem theorem states that all embeddings of a f.d. division algebra in a central simple artinian k-algebra are conjugate, and $E \underset{k}{\circ} k(x)$ has centre k, any embedding of D in $E \underset{k}{\circ} k(x)$ is conjugate to its embedding in E.

We shall restrict our attention for the time being to the coproduct of commutative f.d. extensions of k, and of rational function fields in 1 variable over k. 9.7 shows that any f.d. division subalgebra of such a coproduct must be commutative. We are able to extend 9.9 a little in this context.

<u>Theorem 9.10</u> Let E be a normal extension of k; then any f.d. extension of k in $E \underset{k}{\circ} k(x)$ is conjugate to a subfield of E.

Proof: Let C be a commutative subfield of $E \circ_k k(x)$, $[C:k] < \infty$; then h.c.f. $\{[A_i:E]\} = 1$, where A_i runs through field joins of E and C inside the algebraic closure of k. But since E is normal, there is a unique such field join over k, and so, $[A_i:E] = [A_j:E] = 1$. That is, C embeds in E. Conjugacy follows from the Noether, Skolem theorem as before.

Before we use these results to get further restrictions on the f.d. division subalgebras of skew field coproduct of commutative extensions we need some embedding lemmas for general skew field coproducts. In order to explain the proof cleanly we shall need the next definition.

Consider a tree whose edges are labelled by skew fields and whose vertices are labelled by rings so that if R_v is associated to the vertex v and D_e is the skew field associated to an edge e on which v lies there is a specified embedding of D_e in R_v; the <u>tree coproduct</u> associated to this <u>tree of rings</u> is generated by copies of the vertex rings R_v together with the relations that if ιe and τe are the vertices of the edge e then $R_{\iota e} \cap R_{\tau e} = D_e$; it is easy to see that this tree coproduct may be got by a series of coproducts amalgamating skew fields and so if all the vertex rings are skew fields the tree coproduct must be a fir by Bergman's coproduct theorems.

<u>Lemma 9.11</u> 1/ Let $\{E_i : i = 1 \text{ to } n\}$, $\{F_i : i = 1 \text{ to } n\}$ be families of skew fields such that $E_i \subseteq F_i$; then $\circ_k E_i$ embeds in $\circ_k F_i$.

2/ $\circ_k E$ over a finite indexing set embeds in $E \circ_k k(x)$.

Proof: The ring $\sqcup_k E_i$ embeds in $\sqcup_k F_i$; the universal localisation of $\sqcup_k F_i$ at the full matrices over $\sqcup_k E_i$ is the tree coproduct of rings associated to the tree of rings:

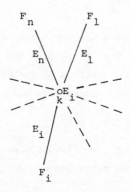

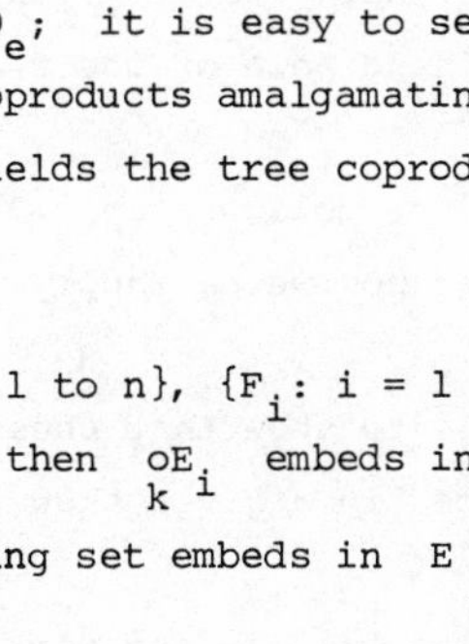

which is a fir since the vertex groups are skew fields; its universal skew field must be $\circ_k F_i$ since this is the only universal localisation of $\circ_k F_i$ that is a skew field.

For the second part, $\circ_k E$ over a finite indexing set embeds in $\circ_k E$ over the countable indexing set $\mathbb{Z}$; this has the automorphism σ of infinite order that shifts the indexing set by 1; we form the skew field $(\circ_k E)(x:\sigma)$ which it is easy to see is isomorphic to $E \circ_k k(x)$.

We can put these results together to find a useful restriction on the f.d. extensions of k lying in the skew field coproduct of f.d. commutative extensions.

<u>Theorem 9.12</u> Let $\{E_i : i = 1 \text{ to } n\}$ be a finite collection of f.d. extensions of k, and let D be a f.d. division subalgebra of $\circ_k E_i$; then D is commutative, its dimension divides $\text{l.c.m.}\{[E_i:k]\}$, and it can be embedded in the normal closure E of a field join of the fields E_i.

Proof: The first two conditions follow from 9.5 and 9.6. We have shown that $\circ_k E_i$ embeds in $E \circ_k k(x)$, and now we can apply 9.10.

We give an example to show that these conditions are not sufficient for a field to embed in a given skew field coproduct.

<u>Example 9.13</u> Consider $\mathbb{Q}(\sqrt{2}) \circ_{\mathbb{Q}} \mathbb{Q}(\sqrt{3})$; the normal closure of $\mathbb{Q}(\sqrt{2},\sqrt{3})$ is itself, and $\mathbb{Q}(\sqrt{6})$ is a subfield satisfying the consequences of 9.11. But $\mathbb{Q}(\sqrt{6}) \otimes_{\mathbb{Q}} \mathbb{Q}(\sqrt{2})$ and $\mathbb{Q}(\sqrt{6}) \otimes_{\mathbb{Q}} \mathbb{Q}(\sqrt{3})$ are both fields, so that by 9.3, $\mathbb{Q}(\sqrt{6})$ does not embed in $\mathbb{Q}(\sqrt{2}) \circ \mathbb{Q}(\sqrt{3})$.

As has become clear, the coproduct of separable extensions does not behave particularly well with respect to the property that f.d. division subalgebras are conjugate to a subfield of one of the factors; it comes as a pleasant surprise that purely inseparable extensions work extremely well in this context.

<u>Theorem 9.14</u> Let $\circ_k E_i$ be the skew field coproduct of the skew fields E_i; if C is a purely inseparable extension of k lying in $\circ_k E_i$, it is

conjugate to a subfield of some E_i.

Proof: Let p be the characteristic of k; we shall show that the dimension over E_i of a simple $C\otimes_k E_i$ module is a power of p, so that C embeds in $\underset{k}{\circ}E_i$ if and only if it embeds in some E_i; as usual, the Noether, Skolem theorem completes the proof.

Let C_i be the centre of E_i; then the centre of $C\otimes_k E_i$ is $C\otimes_k C_i$, which is a local artinian ring, because C is a purely inseparable extension of k, and so, has a unique field join, C', with any field extension, C_i, of k. C' is a purely inseparable extension of C_i, so $[C':C_i]$ is a power of p. $C'\otimes_k E_i$ is the simple artinian image of $C\otimes_k E_i$ modulo its radical, which implies that the dimension of a simple $C\otimes_k E_i$ module over E_i must divide $[C':C_i]$ which is a power of p. We conclude the proof as was indicated previously.

This fits well with earlier results.

<u>Theorem 9.15</u> Let $\underset{k}{\circ}E_i$ be a skew field coproduct of finitely many f.d. purely inseparable extensions of k; then any f.d. subfield of $\underset{k}{\circ}E_i$ is conjugate to a subfield of one of the factors.

Proof: By 9.12, any subfield must lie in the normal closure of a field join of the fields E_i. But this is simply the unique purely inseparable field extension of k generated by the fields E_i. So, any such subfield is purely inseparable, and, by 9.14, it must be conjugate to a subfield of one of the factors.

There is a further situation, where one could hope to prove that a f.d. subfield of a skew field coproduct of commutative fields would be conjugate to a subfield of one of the factors; when each factor has dimension p over k, where p is a prime. This turns out to be false in general, but the set of primes for which it is true is an interesting collection. The next result depends on the classification of finite simple groups, since we use Cameron's classification of doubly transitive permutation groups (Cameron 81).

<u>Theorem 9.16</u> Let p be a prime not equal to 11 or $(q^t-1)/(q-1)$, where q is a prime power, and $t > 2$. Then, if $\{E_i\}$ is a finite collection of field extensions each of dimension p over k, any f.d. subfield of $\underset{k}{\circ}E_i$

is conjugate to one of the factors. For the proscribed primes, there are counter-examples.

Proof: Suppose that E embeds in $\underset{k}{o}E_i$, $[E:k] < \infty$, and that $E \not\cong E_i$ for any i. By 9.5, $[E:k] = p$; and E cannot be purely inseparable by 9.14. So, it is separable.

Since E embeds in $\underset{k}{o}E_i$, there is some i for which $E \otimes E_i \cong \overset{n}{\underset{j=1}{\times}} F_{ij}$, $n \geq 2$. Since $E \not\cong E_i$, $[F_{ij}:E_i] \not\cong 1$ for any j. Because $\Sigma[F_{ij}:E_i] = p$, $\text{h.c.f.}_n\{[F_{ij}:E_i]\} = 1$, so that E lies in $E_i \underset{k}{o} E_i$ which implies that E lies in the normal closure, N, of E_i by 9.10. So, E_i is a separable extension of k. Let G be the Galois group of N over k. Since G acts faithfully on the roots of an irreducible polynomial for a generator of E_i over k, p divides $|G|$, but p^2 does not. Since $[E_i:k] = p = [E:k]$, we see that the subgroups H' and H that fix E_i and E respectively are inconjugate p-complements in G; therefore, by Hall's theorem, G cannot be soluble.

The actions of G on the right cosets of H and of H' define non-isomorphic faithful transitive actions of G on a p-element set. By Burnside's theorem (11) we see that transitive actions of an insoluble group on a p-element set are doubly transitive. Cameron (81) shows that this can happen only when $p = 11$ or is of the form $(q^t-1)(q-1)$ for a prime power q, and $t > 2$.

For $p = 11$, there are two different actions of $PSL(2,11)$ on 11 points; for $p = (q^t-1)/(q-1)$, the actions are given by the action of $PGL(q,t)$ on the points, and dually, on the hyperplanes of projective space.

We have reached a contradiction provided that $p \neq 11$, or $(q^t-1)/(q-1)$ for a prime power q and integer $t > 2$. So we need to examine what actually happens in these cases. Given a group G having two such faithful actions on p-element sets, we have two inconjugate p-complements. We may find a Galois extension of fields $N \supset k$, having Galois group G, and the two p-complements give us two non-isomorphic field extensions of k, both of dimension p over k with isomorphic normal closure. Call them E and E'; then E embeds in $E' \underset{k}{o} E'$, and vice versa. Our first example occurs when $p = 7$.

It is of some interest that at present we are unable to tell whether the skew fields $E \circ_k E$, $E \circ_k E'$ and $E' \circ_k E'$ where E and E' are as described in the last paragraph of the proof above are isomorphic. The invariants that we shall develop in chapter 10 are not good enough to settle this point.

Transcendence in skew fields

One question that arises naturally is whether the coproduct of skew fields having no elements algebraic over k also has this property. We shall present a counter-example to this in the course of this section, as well as giving a counter-example to a conjecture of Cohn and Dicks (80), connected to suitable ideas for transcendence in skew fields. We shall also give the definitions for notions of transcendence that do behave well for the skew field coproduct. First, however, we present a positive result in the direction of our original question.

Theorem 9.17 Let C be a commutative subfield of $\circ_k E_i$ such that $[C:k] = p^n$ for some prime p; then one of the skew fields E_i contains elements algebraic over k.

Proof: We lose nothing by assuming that C is a primitive extension of k, $C = k(\alpha)$.

Let L_i be the centre of E_i; if $C \otimes_k L_i$ is not a field, the irreducible polynomial of α over k splits over L_i, and the coefficients of the factors generate an algebraic extension of k in L_i. So, assume that $C \otimes_k L_i$ is a field; then $C \otimes_k E_i \cong (C \otimes_k L_i) \otimes_{L_i} E_i$ is simple artinian, and so, the dimension of a simple $C \otimes_k E_i$ module over E_i divides $[C:k] = p^n$, and must itself be a power of p. Since C embeds in $\circ_k E_i$, 11.4 shows that the h.c.f. of these numbers is 1, and therefore, there is some i for which the simple $C \otimes_k E_i$ module has dimension 1 over E_i; that is, C embeds in E_i.

Of course, we cannot conclude that C embeds in some E_i even when $[C:k]$ is prime, as we saw in 9.16.

We can re-phrase 9.17 in the form that an algebraic extension of k lying in the skew field coproduct of two skew fields having no algebraic elements must have dimension divisible by at least two primes. We shall find such a subfield of dimension 6.

<u>Example 9.18</u> Let $E \supset k$ be an extension of fields such that $[E:k] = 6$, and there are no intermediate fields; this occurs, for example, if the Galois group of the Galois closure of E over k is S_6. We shall construct a pair of skew fields E_1 and E_2 such that E embeds in $E_1 \circ_k E_2$ although neither E_1 nor E_2 have any elements algebraic over k.

Let E_1 be the skew field such that $M_2(E_1) \cong M_2(k) \circ_k E$, and let E_2 be the skew field such that $M_3(E_2) \cong M_3(k) \circ_k E$. Certainly, E embeds in $E_1 \circ_k E_2$ since there is a simple $E \otimes_k E_1$ of dimension 2 over E_1, and a simple $E \otimes_k E_2$ module of dimension 3 over E_2. Since h.c.f.$\{2,3\} = 1$, it is clear that E embeds in $E_1 \circ_k E_2$. It remains to show that E_1 and E_2 have no algebraic elements over k.

Let C be a commutative subfield of E_1 such that $[C:k] = n$. Since C lies in E_1, and $M_2(E_1) \cong M_2(k) \circ_k E$, there is a simple $M_2(k) \circ_k E$ module of rank $\frac{1}{2}$ over $M_2(k) \circ_k E$, so that by 11.4, $\frac{1}{2} = \text{h.c.f. } \{n/2, [C_j:E]\}$ as C_j runs through simple $C \otimes_k E$ modules. So n must be odd, and the equation $1 = \text{h.c.f. } \{n, [C_j:E]\}$ holds. Since $[C \otimes_k E:E] = n$, n is a sum of the numbers $[C_j:E]$, and so, $1 = \text{h.c.f.}\{[C_j:E]\}$, which is equivalent to the hypothesis that C embeds in $E \circ_k E$ and n is odd. By 9.6, n divides 6, so $n = 1,3$. Since $\text{h.c.f.}\{[C_j:E]\} = 1$, it follows in either case that for some j, $[C_j:E] = 1$, that is, C embeds in E. By assumption, E has no non-trivial subfields, so that $C = k$.

We have shown that E_1 has no non-trivial algebraic extensions of k inside it; an entirely similar argument shows that E_2 has no non-trivial extensions either. Thus $E_1 \circ_k E_2$ is a coproduct of skew fields both of which have no algebraic elements over k except for those in k, but the coproduct does have non-trivial algebraic elements.

We give some variants of the definition of transcendence that do behave well with respect to the coproduct construction, and discuss the connections of these notions with those in (Cohn, Dicks 81).

We define a skew field to be n-transcendental over a central subfield k if for all f.d. division algebras D over k of p.i. degree dividing n, $D \otimes_k E$ is a skew field. We define a skew field to be totally transcendental over k if it is n-transcendental for all n. It is clear that n-transcendental implies that there are no algebraic elements over k that are not in k. The skew fields of example 9.17, E_1 and E_2, are not even 1-transcendental.

Theorem 9.19 The coproduct of n-transcendental skew fields over k is n-transcendental. Therefore, the coproduct of totally transcendental skew fields is totally transcendental.

Proof: The proof is clear from 9.3.

We consider the definitions of Cohn and Dicks. A skew field E is said to be regular over k, if $E\otimes_k K$ is a domain for all commutative fields K over k. This is shown in their paper to be equivalent to $E\otimes_k \bar{k}$ is a skew field where k is the algebraic closure of k. So regular is equivalent to 1-transcendental.

Cohn and Dicks define k to be totally algebraically closed in E if $E\otimes_k L$ is a skew field for all simple algebraic extensions L of k. They ask whether this together with the condition that $E\otimes_k k^{p^{-\infty}}$ is a skew field where p is the characteristic of k, and $k^{p^{-\infty}}$ is the maximal purely inseparable extension of k imply that E is 1-transcendental. We shall find that there are a number of counter-examples to this that arise as a consequence of the next theorem, which is a perhaps more natural result.

Theorem 9.20 Let $F \supset k$ be a normal extension of commutative fields such that $[F:k] = p^n$, where p is a prime. Let D be the skew field such that $M_q(D) \cong M_q(k) \underset{k}{\circ} F$, where $q = p^{n-1}$. Then for any algebraic extension E of k, $D\otimes_k E$ has zero-divisors if and only if $E \supset F$.

Proof: Let $[E:k] = e$; as usual, we shall use 9.3. The rank of a simple $M_q(E)$ module as $M_q(k)$ module is e/q. The simple modules for $E\otimes_k F$ correspond to field joins of E and F, and there is a unique such field join up to isomorphism over k since F is a normal extension of k. We have the following diagram, where the dimension of the extensions is the label of the edges.

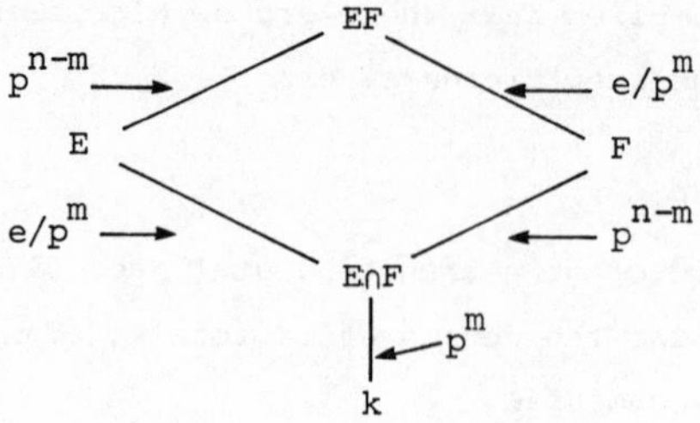

We see that $[EF:F] = e/p^m$ and so the rank of a simple $M_q(D)\otimes_k E$ module is h.c.f. $\{e/q,\ e/p^m\}$. $D\otimes_k E$ has zero-divisors if and only if the rank of a simple $Mq(D)\otimes_k E$ module is less than e/q. Since e/q divides e/p^m or vice versa, this number is less than e/q if and only if e/p^m divides e/q and is not equal to it. But $m \leq n$, and $q = p^{n-1}$, so this happens exactly when $m = n$. In this case, $[EF:k] = e$, so that E contains F.

<u>Theorem 9.21</u> Let $F \supset k$ be a normal extension of k that is neither simple nor purely inseparable, and let $[F:k] = p^n$. Let D be the skew field such that $M_q(D) \cong M_q(k) \underset{k}{\circ} F$, where $q = p^{n-1}$. Then $D\otimes_k k^{p^{-\infty}}$ is a skew field, and k is totally algebraic closed in D, although D is not 1-transcendental over k.

Proof: In 9.20, we showed that $E\otimes_k D$ has zero-divisors for a f.d. extension, E, of k if and only if E contains F. If E contains F, it cannot be a simple extension nor can it purely inseparable. Therefore, $D\otimes_k k^{p^{-\infty}}$ is a skew field, and k is totally algebraically closed in D, but D is not 1-transcendental over k.

We have used skew fields D of the form $M_t(D) \cong M_t(k) \underset{k}{\circ} E$ where E is a f.d. commutative extension of k as our main construction recently. Before leaving them, we mention one further property that has some interest. Let $[E:k] = n$, where E is a simple extension $E = k[\alpha]$. Let t divide n; and let D be the skew field such that $M_t(D) \cong M_t(k) \underset{k}{\circ} E$. Then the dimension of a simple $D\otimes_k E$ module over D is t as one checks from 9.3. That is, the minimal polynomial of over k splits over D into n/t factors of length t.

Generic partial splitting skew fields

Amitsur (55) and Roquette (62) developed the notion of the generic splitting field of a central simple f.d. algebra over a field k. This construction has a natural non-commutative analogue, which we shall develop and study in the course of this section. We shall investigate its f.d. division algebras using the techniques developed earlier for such questions.

Let D be a central division algebra over the field k, $[D:k] = n^2$, and let m divide n^2. A skew field E is said to be an m-splitting skew field of D if $D\otimes_k E \cong M_m(S)$ for some simple artinian ring S. A commutative field can only be an m-splitting field when m divides n; however, D^o is an n^2-splitting skew field of D. There are universal or generic m-splitting fields that are commutative for dividing n, and we intend to show that there are suitable skew analogues of these constructions. We shall construct a ring that represents the functor $\mathrm{Hom}(M_m(k), D\otimes_k -)$ in the category of k-algebras, and then show that it has a universal skew field of fractions, which will be the skew field that we are interested in.

Theorem 9.22 The functor $\mathrm{Hom}(M_m(k), D\otimes_{k}-)$ is representable by the k-algebra R, where $D \sqcup_k M_m(k) \cong D\otimes_k R$. In particular, E is an m-splitting skew field for D if and only if $\mathrm{Hom}(R,E)$ is not empty.

Proof: Let $\alpha \in \mathrm{Hom}(M(k)), D\otimes_k A)$ for some k-algebra A. Then $1 \sqcup \alpha : D \sqcup_k M_m(k) \to D\otimes_k A$ is a homomorphism and must take the centraliser of D in the first ring to A, which gives us a map from R to A.

Conversely, given $\beta: R \to A$, we have a map $1 \otimes\beta: D\otimes_k R \to D\otimes_m A$, which induces a map $\bar{\beta}: M_m(k) \to M_m(k) \sqcup_k D \cong D\otimes_k R \xrightarrow{\beta} D\otimes_k A$.

The two processes above are mutually inverse, so R represents the functor as we wished to show. The last sentence is trivial.

We shall look at R a little more closely, but first we tighten up the notation. The ring representing the functor $\mathrm{Hom}(M_m(k), D\otimes_k -)$ is denoted by $R(D,m)$.

Theorem 9.23 $R(D,m)$ is an hereditary domain whose monoid of f.g. projectives is generated by the free module of rank 1, and a projective P satisfying the relation $R^{n^2} \cong P^m$, and no other relations. Therefore, it has a unique rank function on its f.g. projectives, and this rank function takes values in $\mathbb{Z}$. So, it has a universal skew field of fractions.

Proof: $D\otimes_k R \cong D \sqcup_k M_m(k)$, so

$M_{n^2}(R) \cong D^o\otimes_k (D\otimes_k R) \cong D^o\otimes_k (D \sqcup_k M_m(k)) \cong M_{n^2}(k) \sqcup_{D^o} M_m(D^o)$, where in this isomorphism R is the centraliser of the first factor.

By Bergman's coproduct theorems, 2.18, R is a domain and all f.g. projectives are induced up from the factors in the coproduct representation $M_{n^2}(R) \cong M_{n^2}(k) \sqcup_{D^o} M_m(D^o)$. Morita equivalence shows that the monoid of f.g. projectives is as stated, and R is hereditary.

We have shown all but the last sentence. Since $P^m \cong R^{n^2}$, its rank under any rank function can only be n^2/m, so that there is a unique rank function, and the universal localisation of R at this rank function is a skew field.

Let $U(D,m)$ be the universal localisation of $R(D,m)$ at the unique rank function. We shall call it the <u>universal m-splitting skew field of D</u>. We wish to investigate the f.d. division algebras over k that embed in it.

<u>Theorem 9.24</u> Let $U(D,m)$ be the universal m-splitting skew field of D; then the f.d. division subalgebras of $U(D,m)$ are isomorphic to the f.d. division algebras E_i of D^o such that h.c.f.$\{[E_i:k],n^2/m\} = 1$. In particular, if p divides m implies that p divides n^2/m for all primes p, $U(D,m)$ has no elements algebraic over k except for those in k.

Proof: We recall from 9.23 that $U = U(D,m)$ is the universal skew field of fractions of R, where $M_{n^2}(R) \cong M_{n^2}(k) \sqcup_{D^o} M_m(D^o)$. Hence,

$M_{n^2}(U) \cong M_{n^2}(k) \circ_{D^o} M_m(D^o)$. By 9.3, the rank of a simple $E^o\otimes_k M_{n^2}(U)$ module over $M_{n^2}(U)$ is given by h.c.f.$\{[E:k]/n^2,[S:D^o]/m\}$ where S is the unique simple $E^o\otimes_k D^o$ module. So, $[S:D^o]$ divides $[E:k]$.

If E embeds in U, we know that this number is $1/n^2$, which happens if and only if $[S:D^o] = 1$, and h.c.f.$\{[E:k]/n^2,1/m\} = 1/n^2$, which is equivalent to the two statements:

1/ E embeds in D^{o}, and

2/ h.c.f.$\{[E:k],n^2/m\} = 1$.

This proves all but the last sentence, which is an easy corollary. It occurs for example, if $n = m$.

Division subalgebras of the universal skew fields for rings with weak algorithm

This final section is rather technical; we use 9.2 to describe the f.d. division subalgebras of the universal skew field of fractions of a ring with weak algorithm. For the record, we define what a ring with weak algorithm is; however, the reader is more likely to gain some understanding of the notion by reading chapter 2 of (Cohn 71). Rings with weak algorithm are a generalisation of tensor rings of bimodules over skew fields.

Let R be a ring with a filtration over $\mathbb{N}$; that is, we have a function v from R to $\mathbb{N}$ such that $v(x) \geq 0$ for $x \neq 0$; $v(0) = -\infty$; $v(x-y) \leq \max\{v(x), v(y)\}$; $v(xy) \leq v(x) + v(y)$; $v(1) = 0$.

Let $\{a_i : i \in I\}$ be a subset of the elements of R; we say that it is right dependent with respect to the filtration v if there exist elements $\{b_i : i \in I\}$ that are almost all 0, such that $v(\Sigma a_i b_i) < \max_i \{v(a_i) + v(b_i)\}$ or else, some a_i is 0.

We say that an element a of R is right dependent with respect to v on the set $\{a_i\}$ if there exist $\{b_i : i \in I\}$ almost all 0 such that $v(a - \Sigma a_i b_i) < v(a)$ whilst $v(a_i) + v(b_i) \leq v(a)$ for each i.

We say that R satisfies the weak algorithm with respect to the filtration v, if given any set of elements $\{a_i : i = 1 \text{ to } n\}$ right dependent with respect to v such that $v(a_i) \leq v(a_j)$ for $i < j$, then some a_i is right dependent with respect to v on the preceding set $\{a_1, a_2 \ldots a_{i-1}\}$.

We shall assume the results of chapter 2 in (Cohn 71) throughout this section. In particular, R is a fir whose universal skew field of fractions we call U. R_0, the set of elements of whose filtration is 0 together with 0 forms a skew field, and R is a two-sided Ore domain only when it takes the form $R = R_0[x;\alpha,\delta]$, where α is an automorphism and δ is a $(1,\alpha)$-derivation. When R is not a two-sided Ore domain, the centre of U lies in R_0 by 8.1, since R_0 is the group of units of R. When $R = R_0[x;\alpha,\delta]$ we shall need to use different techniques which we present at the end of the section. In the case where R is not two-sided Ore, let K be the centre both of R and U.

Theorem 9.25 Let R be a ring with weak algorithm with universal skew field U and 0th term in its filtration R_0; let D be a f.d. division K-algebra. Then the dimension over U of a simple $D^{\circ} \otimes_K U$ module is equal to h.c.f. $\{[N_i:R_0]\}$ as N_i runs through the f.g. $D^{\circ} \otimes_K R^{\circ}$ modules.

Proof: By 9.2, the dimension of a simple $D^{\circ} \otimes_K U$ module is given by h.c.f. $\{[M_j:R_0]\}$ as M_j runs through f.g. $D^{\circ} \otimes_K R$ submodules of free modules. For ease of notation, we shall consider left D, right R bimodules on K-centralising generators, since these are clearly the same thing.

We recall that R is filtered by

$$R_0 \subset R_1 \subset R_2 \subset \ldots\ldots \subset R, \quad \cup R_i = R.$$

Let F be a free left D, right R bimodule on the set X, which we filter by

$$DXR_0 \subset DXR_1 \subset DXR_2 \subset \ldots\ldots \subset DXR = F.$$

For $f \in F$, we define $v(f) = \min\{r: f \in DXR_r\}$. Let M be a f.g. left D, right R sub-bimodule of F. We construct a basis for M as an R-module from which it will be clear that the dimension of M as an R-module is a sum of the numbers $[N_i:R_0]$ for D, R_0 bimodules N_i.

We choose our basis by induction. Let $M_0 = 0$. Suppose that at the ith stage we have a D, R submodule of M generated as R module by elements $\{a_j\}$ lying in DXR_n such that $\{a_j\}$ is a basis of this module, M_{i-1}, and no element of $M - M_{i-1}$ lies in DXR_n, where DXR_n is minimal subject to containing a_j.

Let $A_i = \{a: a \in M - M_{i-1}, v(a)$ is minimal$\}$. Since M_{i-1} and M are invariant under the action of D, it is clear that $A_i + M_{i-1}/M_{i-1}$ is a left D, right R_0 bimodule; moreover, it is a consequence of the weak algorithm that if $M_i = A_i R + M_{i-1}$, a basis for $A_i + M_{i-1}/M_{i-1}$ over R_0 together with the set $\{a_j\}$ is a basis for M_i as R module.

Since M is f.g., $M = M_m$ for some m, and $[M_i:R]$ is finite for all i. But it is clear that $[M_{i+1}:R] - [M_i:R]$ is equal to the dimension of some D, R_0 bimodule. So, we see that the dimension of a simple D, U bimodule is equal to the h.c.f.$\{[M:R]\}$ for f.g. left D, right R bimodules, which must be divisible by h.c.f.$\{[N_i:R_0]\}$ for f.g. D, R_0

bimodules by our argument. However, given a D, R_0 bimodule N, we construct the D, U bimodule $N\otimes_{R_0} U$. It is clear that $[N\otimes_{R_0} U:U] = [N:R_0]$, and so, the dimension over U of a simple D, U bimodule must divide h.c.f.$\{[N_i:R_0]\}$ which forces equality.

To complete this section, we shall deal with the two-sided Ore case. We recall that in this case, $R = R_0[x;\alpha,\delta]$ where R_0 is a skew field, α is an automorphism and δ is a $(1,\alpha)$ derivation.

<u>Theorem 9.26</u> Let U be the universal skew field of fractions of $R_0[x;\alpha,\delta]$. Let K be the intersection of the centre of U with R_0. Then any f.d. division K-algebra lying in U is isomorphic to a skew subfield of R_0.

Proof: We shall show that there is a valuation on U trivial on R_0 whose residue class skew field is identified with R_0. If D is a f.d. division K-algebra in U, the valuation must be trivial on D since it is a f.d. extension of K, and so, the valuation induces an embedding of D in R_0.

The construction of the valuation essentially occurs on p.18 of Cohn (77). We summarise it briefly here.

Set $y = x^{-1}$ and write out the commutation formula:
$rx = xr^{\alpha} + r^{\delta}$, so $yr = r^{\alpha}y + yr^{\delta}y$.

The set of power series over R_0 in y with co-efficients on the left and the stated commutation rule (which allows us to re-write any expression in R_0 and y as a power series of the given form) is a principal valuation domain; the ring $R_0[x;\alpha,\delta]$ embeds in its skew field of fractions by $x \to y^{-1}$. The residue class skew field of the valuation is R_0, and the valuation is trivial on R_0 as we wished, so our proof is complete.

10 THE UNIVERSAL BIMODULE OF DERIVATIONS

When studying a homomorphism of commutative rings $\phi:R \to S$, it is often useful to look at the module of relative differentials of S over R. There is a non-commutative analogue of this construction, the universal bimodule of derivations, which, in many situations of interest to us here is very powerful. It will enable us to find useful numerical invariants of certain skew field coproducts, that allow us to distinguish between some of them; in particular, we shall be able to distinguish between free skew fields on different numbers of generators. On the way, we shall be able to characterise those epimorphisms from an hereditary k-algebra to skew fields that arise as universal localisations by the associated map on the bimodule of derivations over k.

The results in the first section of this chapter from 10.4 to 10.6 that calculate the universal bimodule of derivations for certain universal constructions are all from the work of Bergman and Dicks (75,78).

Calculating the universal bimodule of derivations

In the commutative case, we look at derivations from the commutative ring to modules over it, that vanish on a given subring; the natural non-commutative generalisation of this is to look at derivations to bimodules over the ring; if R is an R_0-ring we are interested in derivations that vanish on the image of R_0. On general principles there is a universal such derivation $\delta:R \to \Omega_{R_0}(R)$; that is, the functor $\mathrm{Der}_{R_0}(R,M)$ which associates to a bimodule M the set of derivations from R to M that vanish on R_0 is naturally isomorphic to the functor $\mathrm{Hom}_{R\text{-bim.}}(\Omega_{R_0}(R),M)$ where the derivation associated to a particular homomorphism $\alpha:\Omega_{R_0}(R) \to M$ is the composite $\delta\alpha$. We can construct the bimodule in the following way: for each element r of R, we have a generator δr, and we impose the relations $\delta s = 0$, for s in the image of R_0, $\delta(r_1 + r_2) = \delta(r_1) + \delta(r_2)$, and

$\delta(r_1r_2) = r_1\delta(r_2) + \delta(r_1)r_2$. It is clear that the map $\delta: R \to \Omega_{R_O}(R)$ has the desired universal property; however, the nature of this bimodule is opaque from this description, so we shall first give an alternative description, and then we shall use it to simplify our presentation above. Our new description gives us a useful connection between the universal bimodule of derivations and homology.

<u>Theorem 10.1</u> Let R be an R_O-ring; then there is an exact sequence:

$$O \to \Omega_{R_O}(R) \to R\otimes_{R_O} R \overset{m}{\to} R \to O,$$

where m is the multiplication map, and the universal derivation in this representation is given by $\delta(x) = x\otimes 1 - 1\otimes x$.

Proof: Certainly, the map from R to $\ker(m)$ given by $d: x \to x\otimes 1 - 1\otimes x$ is a derivation vanishing on R_O. So the composite of this derivation with any bimodule map gives a natural transformation from $\mathrm{Hom}_{R\text{-bimod}}(\ker(m), \_)$ to the functor $\mathrm{Der}_{R_O}(R, \_)$, which must be injective since the kernel of m is generated by the elements $x\otimes 1 - 1\otimes x$.

Given any derivation vanishing on R_O, $d': R \to M$ where M is an R,R bimodule, we define a bimodule homomorphism from $\ker(m)$ to M by the formula $\alpha_{d'}(\Sigma x_i\otimes y_i) = \Sigma d'(x_i)y_i = -\Sigma x_i d'(y_i)$, which works because d' is a derivation and $\Sigma x_i y_i = 0$. From these equations, it is clear that this map is a bimodule homomorphism, and that $d' = d\alpha_{d'}$; so the derivation $d: R \to \ker(m)$ has the correct universal property, and by the uniqueness of an object representing a functor, we see that our theorem holds.

We can simplify our previous description a little using this result. From the relations, $\delta(r_1r_2) = r_1\delta(r_2) + \delta(r_1)r_2$ and $\delta(r_1 + r_2) = \delta(r_1) + \delta(r_2)$, it is clear that if R is generated over R_O by the set of elements X, then $\Omega_{R_O}(R)$ is generated by $\delta(X)$. We wish to determine the relations imposed, so first we consider the case where the set X is a free generating set.

<u>Theorem 10.2</u> Let $R \cong R_O \sqcup_{\mathbb{Z}} \mathbb{Z}\langle X\rangle$, the ring generated over R_O by the set X subject to no relations; then $\Omega_{R_O}(R)$ is the free bimodule over R on

the set X.

Proof: Let M be any R,R bimodule; then any derivation from R to M vanishing on R_0 is determined by the image of the elements X under the derivation; moreover, the elements of X may be mapped anywhere by a derivation. So, the functor $\mathrm{Der}_{R_0}(R,M)$ is naturally equivalent to the functor $\mathrm{Hom}_{\mathrm{Sets}}(X,M)$; one object that represents this functor is the free bimodule over R on the set δX, which has a derivation from R to it extending the map $X \to \delta X$; by the uniqueness of an object representing a functor, $\Omega_{R_0}(R)$ must be isomorphic to this bimodule and the universal derivation from R to this free bimodule is the one mentioned earlier.

Given an arbitrary R_0-ring, R, generated over R_0 by a set of elements X, there is a surjection $R_0 \underset{\mathbb{Z}}{\sqcup} \mathbb{Z}\langle X\rangle \to R$; so we should like to describe how the universal bimodule of derivations changes under surjective homomorphisms between R_0-rings. This is quite simple to describe as we shall see in the next theorem.

First of all, we introduce some notation, which simplifies the equations of this chapter substantially. In many situations, we shall have a specific homomorphism of rings $R \to S$ and also a particular R,R bimodule M; we may form the S,S bimodule $S\otimes_R M\otimes_R S$, which we shall in general write as ${}^{\otimes}M^{\otimes}$; if the rings R and S are constructed from other rings in a manner which is reflected in the names of R and S, $S\otimes_R M\otimes_R S$ will be quite unwieldy. For example, we shall prove later on the following formula:

$$\Omega_{R_0}(R_1 \underset{R_0}{\sqcup} R_2) \cong ({}^{\otimes}\Omega_{R_0}(R_1)^{\otimes}) \oplus ({}^{\otimes}\Omega_{R_0}(R_2)^{\otimes}) \quad .$$

Since $\Omega_{R_0}(R_1 \underset{R_0}{\sqcup} R_2)$ must be an $R_1 \underset{R_0}{\sqcup} R_2$, $R_1 \underset{R_0}{\sqcup} R_2$ bimodule, whilst $\Omega_{R_0}(R_i)$ is an R_i,R_i bimodule, it is clear that by the symbol ${}^{\otimes}\Omega_{R_0}(R_i)^{\otimes}$ we mean $(R_1 \underset{R_0}{\sqcup} R_2)\otimes_{R_i}(\Omega_{R_0}(R_i))\otimes_{R_i}(R_1 \underset{R_0}{\sqcup} R_2)$.

<u>Theorem 10.3</u> Let R be an R_0-ring, and I an ideal of R; then we have an exact sequence: $I/I^2 \to {}^{\otimes}\Omega_{R_0}(R)^{\otimes} \xrightarrow{p} \Omega_{R_0}(R/I) \to 0$, where p is induced by the ring homomorphism, and the map from I/I^2 to the bimodule ${}^{\otimes}\Omega_{R_0}(R)^{\otimes}$ is induced by δ.

If R_O is a semisimple artinian ring, our sequence may be extended to an exact sequence: $O \to I/I^2 \to {}^{\otimes}\Omega_{R_I}(R)^{\otimes} \to \Omega_{R_O}(R/I) \to O$.

Proof: We have the exact sequence: $O \to \Omega_{R_O}(R) \to R\otimes_{R_O} R \to R \to O$, which must be split exact as a sequence of left R modules; consequently, we have the exact sequence:

$$O \to R/I\otimes_{R_O}(\Omega_{R_O}(R)) \to R/I\otimes_{R_O} R \to R/I \to O.$$

We look at part of the long exact sequence of $\mathrm{Tor}^R(\ ,R/I)$:
$\mathrm{Tor}_1^R(R/I\otimes_{R_O} R,R/I) \to \mathrm{Tor}_1^R(R/I,R/I) \to {}^{\otimes}\Omega_{R_O}(R)^{\otimes} \to R/I\otimes_{R_O} R/I \to R/I \to O$, where all terms are $R/I,R/I$ bimodules.

It is well-known (and we shall see this soon) that $\mathrm{Tor}_1^R(R/I,R/I) \cong I/I^2$. The map $R/I\otimes_{R_O} R/I \to R/I$ is the multiplication map, so its kernel is isomorphic to $\Omega_{R_O}(R/I)$ by 7.2, which gives us the exact sequence: $I/I^2 \to {}^{\otimes}\Omega_{R_O}(R)^{\otimes} \to \Omega_{R_O}(R/I) \to O$; if R_O is semisimple artinian, $R/I\otimes_{R_O} R$ is projective, so that $\mathrm{Tor}_1^R(R/I\otimes_{R_O} R,R/I) = O$, and we have the exact sequence: $O \to I/I^2 \to {}^{\otimes}\Omega_{R_O}(R)^{\otimes} \to \Omega_{R_O}(R/I) \to O$ as stated.

We need to show that the map from I/I^2 to ${}^{\otimes}\Omega_{R_O}(R)^{\otimes}$ is the one we said; in order to show this, we describe the isomorphism between I/I^2 and $\mathrm{Tor}_I^R(R/I,R/I)$. Consider the map of exact sequences:

$$\begin{array}{ccccccccc}
O & \longrightarrow & I & \longrightarrow & R & \longrightarrow & R/I & \longrightarrow & O \\
 & & \downarrow{\scriptstyle 1\otimes\delta} & & \downarrow & & \| & & \\
O & \longrightarrow & R/I\otimes_R\Omega_{R_O}(R) & \longrightarrow & R/I\otimes_{R_O} R & \longrightarrow & R/I & \longrightarrow & O
\end{array}$$

Applying $\mathrm{Tor}^R(\ ,R/I)$ to it gives us a commutative diagram:

$$\begin{array}{ccccc}
O \to \mathrm{Tor}_1^R(R/I,R/I) & \to & I\otimes_R R/I = I/I^2 & \to & R/I \\
\| & & \downarrow & & \\
\mathrm{Tor}_1^R(R/I,R/I) & \to & {}^{\otimes}\Omega_{R_O}(R)^{\otimes} & &
\end{array}$$

which shows that $I/I^2 \cong \mathrm{Tor}_1^R(R/I,R/I)$, whilst the right hand column demonstrates that the map from I/I^2 to ${}^{\otimes}\Omega_{R_O}(R)^{\otimes}$ is what we want it to be.

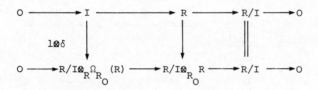

This result gives us a presentation of $\Omega_{R_0}(R)$ in terms of generators and relations whenever we have a presentation of R over R_0. If the set X generates R over R_0, subject to the relations $\{f_i\}$ the universal bimodule of derivations of R over R_0 is generated as a bimodule by the elements δX subject to the relations $\delta f_i = 0$, where δf_i is the formal differential of the element f_i of the ring $R_0 \underset{\mathbb{Z}}{\sqcup} \mathbb{Z}\langle X\rangle$.

We need a few results about the behaviour of the universal bimodule of derivations under various universal constructions; we begin with the coproduct construction.

<u>Theorem 10.4</u> Let $\{R_i : i \in I\}$ be a family of R_0-rings; then

$$\Omega_{R_0}(\sqcup_{R_0} R_i) \cong \oplus ({}^{\otimes}\Omega_{R_0}(R_i)^{\otimes}) .$$

Proof: This is most easily seen using our generator and relation construction of the universal bimodule of derivations described after theorem 7.3. It is clear that $\Omega_{R_0}(\sqcup_{R_0} R_i)$ is generated as a bimodule by the image of the bimodules $\Omega_{R_0}(R_i)$ and there are no further relations, so the theorem must hold.

We can also calculate the universal bimodule of derivations of a tensor ring.

<u>Theorem 10.5</u> Let M be an R,R bimodule; then $\Omega_R(R\langle M\rangle) \cong {}^{\otimes}M^{\otimes}$, where the derivation induces the identity map on M.

Proof: We look at the generator and relation construction of the universal bimodule of derivations again. $\Omega_R(R\langle M\rangle)$ is generated by $\delta(M)$, and the only relations are given by $\delta(rm) = r\delta(m)$, and $\delta(mr) = \delta(m)r$ for r in R, and m in M. So the result follows.

We also wish to study how the universal bimodule of derivations behaves under the process of adjoining universal inverses.

<u>Theorem 10.6</u> Let Σ be a collection of maps between f.g. projectives over the R_0-ring R; then $\Omega_{R_0}(R_\Sigma) \cong {}^{\otimes}\Omega_{R_0}(R)^{\otimes}$.

Proof: Given a bimodule M over a ring T, we can form the ring whose additive structure is isomorphic to $T \oplus M$ and whose multiplication is given by $(t,m)(t',m') = (tt', tm' + mt')$, the trivial extension of T by M; we write this ring as (T,M).

We wish to construct a derivation from R_Σ to ${}^{\otimes}\Omega_{R_O}(R)^{\otimes}$ that extends the universal derivation from R to $\Omega_{R_O}(R)$. So, consider the ring homomorphisms:

$$\begin{array}{ccc} R & \longrightarrow & (R, \Omega R_O(R)) \\ \downarrow & & \downarrow \\ R_\Sigma & & (R_\Sigma, {}^{\otimes}\Omega_{R_O}(R)^{\otimes}) \end{array} \quad ;$$

we wish to complete this to a commutative diagram of ring homomorphisms by a map from R_Σ to $(R_\Sigma, {}^{\otimes}\Omega_{R_O}(R)^{\otimes})$, because such a map must take the form $s \to (s,ds)$ where d is a derivation extending the universal derivation from R to $\Omega_{R_O}(R)$ by the commutativity. However, the set of maps Σ are invertible over $(R_\Sigma, {}^{\otimes}\Omega_{R_O}(R)^{\otimes})$, since they are invertible modulo the nilpotent ideal $(O, {}^{\otimes}\Omega_{R_O}(R)^{\otimes})$; therefore there is a unique map from R_Σ to $(R_\Sigma, {}^{\otimes}\Omega_{R_O}(R)^{\otimes})$ completing the diagram.

We wish to show that this must be a universal derivation. Given a derivation $d':R \to M$ where M is an R_Σ, R_Σ bimodule such that d' vanishes on R_O, we know from the universal property of $\Omega_{R_O}(R)$ that d' restricted to R factors through $\Omega_{R_O}(R)$, so we have a diagram:

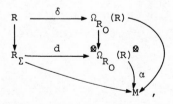

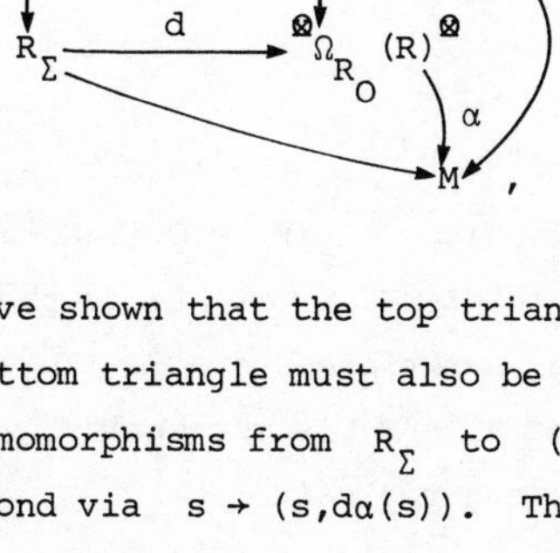

,

where we have shown that the top triangle is commutative and we wish to show that the bottom triangle must also be commutative to complete the proof. We have two homomorphisms from R_Σ to (R_Σ, M) the first via $s \to (s,d'(s))$, and the second via $s \to (s,d\alpha(s))$. These agree on R, and since $R \to R_\Sigma$

is an epimorphism they must be the same map.

Generators for the free skew field

For a free group on n generators it is well known and easy to see that any generating set of n elements must be a free generating set, and the corresponding result for a free k-algebra was shown by Cohn and independently by Lewin (69). It is natural to ask whether the corresponding result holds for free skew fields, which arise as the universal localisation of the free k-algebra; J.Wilson asked whether the free skew field on n generators could be isomorphic to the free skew field on m generators, and a negative answer to this question would follow from the first result. We should also like to know that if D is a skew subfield of E, then the skew subfield of $E\langle X\rangle$ generated by D and X is isomorphic in the natural way to $D\langle X\rangle$, and if M is an E subbimodule of N, then the skew subfield of $E\langle N\rangle$ generated by E and M is naturally isomorphic to $E\langle M\rangle$.

In the first of our problems we shall show that if n elements $t_1,\dots t_n$ generate $k\langle x_1\dots\dots x_n\rangle$ as a skew field over k, the homomorphism ϕ on $k<y_1\dots\dots y_n>$ given by $\phi(y_i) = t_i$ induces an isomorphism:

$$\delta\phi: {}^{\otimes}\Omega_k(k<y_1\dots\dots y_n>)^{\otimes} \cong \Omega_k(k\langle x_1\dots\dots x_n\rangle).$$

It follows from our next theorem that ϕ must extend to an isomorphism of the skew field $k\langle y_1\dots y_n\rangle$ with $k\langle x_1\dots\dots x_n\rangle$ as we wished to show.

Our next result gives us a way of recognising those epimorphisms from a right hereditary k-algebra to a skew field that are universal localisations.

Theorem 10.7 Let E be a skew field, and let R be a right hereditary E-ring. The ring homomorphism $\phi:R \to F$ from R to a skew field F is a universal localisation if and only if ϕ induces an isomorphism $\delta\phi: {}^{\otimes}\Omega_E(R)^{\otimes} \cong \Omega_E(F)$.

Proof: If F is a universal localisation of R, the result is clear. Conversely, if the map $\delta\phi$ is an isomorphism, $\Omega_R(F) = 0$, and so ϕ is at least an epimorphism. Therefore, by theorem 7.5, if Σ is the collection of square matrices over R that become invertible over F, they form a prime matrix ideal, and the induced map $R_\Sigma \to F$ is surjective, where R_Σ is a local ring, and the kernel must be the radical.

Since R_Σ is a universal localisation of a right hereditary ring, it is right hereditary by 4.2, and the kernel of $R_\Sigma \to F$, I, must be a free module, so that $I \neq I^2$ except when $I = 0$.

From 7.3, we have the exact sequence:
$0 \to I/I^2 \to {}^{\otimes}\Omega_E(R_\Sigma)^{\otimes} \to \Omega_E(F) \to 0$; also we have the commutative diagram

$$\begin{array}{ccc} {}^{\otimes}\Omega_E(R)^{\otimes} & \longrightarrow & \Omega_E(F) \\ & \searrow \qquad \nearrow & \\ & {}^{\otimes}\Omega_E(R_\Sigma)^{\otimes}, & \end{array}$$

where the top row and the first slanting arrow are isomorphisms which implies that the second slanting arrow must also be an isomorphism, and $I/I^2 = 0$ follows from the exact sequence above. Hence, $I = 0$, and $R_\Sigma \cong F$.

Next, we show that the skew subfield of $E\langle X\rangle$ generated by D and X where D is a skew subfield of E is isomorphic to $D\langle X\rangle$; the proof is simply to apply the last result.

<u>Theorem 10.8</u> Let D be a skew subfield of E; then the skew subfield of $E\langle X\rangle$ generated by D and X is isomorphic to $D\langle X\rangle$.

Proof: First, we calculate the universal bimodule of derivations of $E\langle X\rangle$ over E. $E\langle X\rangle$ is the universal localisation of the fir $E\langle X\rangle$, so, by 10.6, $\Omega_E(E\langle X\rangle) \cong {}^{\otimes}\Omega_E(E<X>)^{\otimes}$. By the remarks after 10.3, $\Omega_E(E<X>)$ is generated by the elements δX subject to the relations $e\delta x = \delta x e$ for all x in X, and e in E; that is, $\Omega_E(E\langle X\rangle)$ is the free bimodule on the E-centralising generators δX. Let K be the skew subfield of $E\langle X\rangle$ generated by D and X, and consider the K,K bimodule generated by δX in $\Omega_E(E\langle X\rangle)$. This is the image of $\Omega_D(K)$ in $\Omega_E(E\langle X\rangle)$ induced by the ring homomorphism from K to $E\langle X\rangle$. This bimodule is just the free bimodule on the D-centralising generators δX. However, $\Omega_D(D<X>)$ is the free bimodule on the D-centralising set δX, and so, the natural map from $D<X>$ to K induces an isomorphism $\Omega_D(K) \cong {}^{\otimes}\Omega_D(D<X>)^{\otimes}$; since $D<X>$ is a fir, 10.7 shows that K must be isomorphic in the natural way to $D\langle X\rangle$.

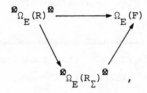

This result gives us a useful handle on our first problem about the generators of free skew fields. Since we are looking at bimodules, it is likely that we shall have to think a little about the <u>enveloping algebra</u> of a k-algebra, R, which is defined as $R^{o} \otimes_k R$; our last result shows that the enveloping algebra of a free skew field embeds in a skew field, and so, the number of generators of a free bimodule over a free skew field is an invariant. In fact, we shall consider a more general result than simply generators for free skew fields, since the method of proof is no harder.

<u>Theorem 10.9</u> Let E be a skew field with central subfield k. Let $t_1 \ldots t_n$ be elements of $E \circ_k k\langle x_1 \ldots x_n \rangle$ that generate it as a skew field over E. Then, if the enveloping algebra of $E \circ_k k\langle x_1 \ldots x_n \rangle$ over k is weakly finite, the natural map from $E \sqcup_k k<y_1 \ldots y_n>$ to $E \circ_k k\langle x_1 \ldots x_n \rangle$ sending y_i to t_i extends to an isomorphism of $E \circ_k k\langle y_1 \ldots y_n \rangle$ with $E \circ_k k\langle x_1 \ldots x_n \rangle$.

Proof: The universal bimodule of derivations of $E \sqcup_k k<x_1 \ldots x_n>$ over E is the free k-centralising bimodule on the generators δx_i, so, by 10.6, the same holds for the universal bimodule of derivations of $E \circ_k k\langle x_1 \ldots x_n \rangle$ over E.

Since $t_1 \ldots t_n$ generate $E \circ_k k\langle x_1 \ldots x_n \rangle$ as a skew field over E, δt_i generate $\Omega_k(E \circ_k k\langle x_1 \ldots x_n \rangle)$. So they are free generators because we assumed that the enveloping algebra of $E \circ_k k\langle x_1 \ldots x_n \rangle$ was weakly finite. Hence, the map from $E \sqcup_k k<y_1 \ldots y_n>$ to $E \circ_k k\langle x_1 \ldots x_n \rangle$ given by $y_i \to t_i$ induces an isomorphism ${}^{\otimes}\Omega_E(E \sqcup_k k<y_1 \ldots y_n>)^{\otimes} \to \Omega_E(E \circ_k k\langle x_1 \ldots x_n \rangle)$.

So, by 10.7, the map sending y_i to t_i extends to an isomorphism from $E \circ_k k\langle y_1 \ldots y_n \rangle$ to $E \circ_k k\langle x_1 \ldots x_n \rangle$.

No example is known of a tensor product of skew fields that is not weakly finite, so that it is yet possible that 10.9 may apply for all skew fields. We shall prove only that the enveloping algebra of $E \circ_k k\langle x_1 \ldots x_n \rangle$ is weakly finite when E is finite-dimensional over k. We shall prove this by embedding $E \circ_k k\langle x_1 \ldots x_n \rangle$ in $M_m(k\langle X \rangle)$ for a suitable set X and natural number m. It is useful to have a more general result on embeddings of simple artinian rings.

<u>Lemma 10.10</u> Let $\{E_i$ for $i = 1$ to $n\}$ be f.d. division algebras over k, such that $\text{l.c.m.}\{[E_i:k]\} = m$; then $\circ_k E_i \circ_k k\langle X \rangle$ embeds in $M_m(k\langle Z \rangle)$ for a suitable set Z.

Proof: E_i embeds in $M_m(k)$, so, by 9.11, $\circ_k E_i$ embeds in $\circ_k M_m(k)$ for n copies of $M_m(k)$, which embeds in $\circ_k M_m(k)$ for countably many copies of $M_m(k)$, which we index by $\mathbb{Z}$. We have an automorphism, σ, of this defined by sending the ith copy of $M_m(k)$ to the $(i + 1)$st, so we form the skew Laurent polynomial ring $(\circ_k M_m(k))[x,x^{-1};\sigma]$, which is a prime principal ideal ring, and so, has a simple artinian ring of fractions. However, we may construct this ring as a universal localisation of $M_m(k) \sqcup_k k[y]$. Form the ring $M_m(k) \sqcup_k k[y,y^{-1}]$, which is clearly a universal localisation of $M_m(k) \sqcup_k k[y]$, and consider the subring generated by the conjugates of $M_m(k)$ by the powers of y; there can be no relations between the copies of $M_m(k)$, so this ring is just $\sqcup_k M_m(k)$ for countably many copies of $M_m(k)$. We adjoin the universal inverses of all full maps between f.g. projectives over $\sqcup_k M_m(k)$ inside $M_m(k) \sqcup_k k[y,y^{-1}]$, and the ring we find must be our skew Laurent polynomial ring above. The simple artinian ring of fractions of this ring must be a universal localisation of $M_m(k) \sqcup_k k[y]$, so it can only be $M_m(k) \circ_k k(y)$.

So far, we have embedded $\circ_k E_i \circ_k k\langle X\rangle$ in the simple artinian ring $M_m(k) \circ_k k(y) \circ_k k\langle X\rangle$, which is just $M_m(k) \circ_k k\langle X \cup y\rangle$. This is a universal localisation of $M_m(k) \sqcup_k k\langle X'\rangle$ for $X' = X \cup y$. However, this is just $M_m(k\langle Z\rangle)$ where Z is the set $\{z_{ijx} : i,j = 1 \text{ to } m,\ x \in X'\}$ where the isomorphism sends x in X' to the matrix $\{z_{ijx}\}$ whose ijth entry is z_{ijx}. Therefore, $M_m(k) \circ_k k\langle X'\rangle$ is isomorphic to $M_m(k\langle Z\rangle)$.

We can prove the theorem we wanted now.

Theorem 10.11 Let E be a f.d. division algebra over k; then n elements of the skew field $E \circ_k k\langle x_1 \ldots . x_n\rangle$ that generate it as a skew field over E are free generators.

Proof: By the last lemma, $E \circ_k k\langle x_1 \ldots . x_n\rangle$ embeds in $M_m(k\langle Z\rangle)$ for some set Z. Consequently, its enveloping algebra over k embeds in $M_{m^2}(k\langle Z\rangle^{\circ} \otimes_k k(Z))$. So we need to show that the enveloping algebra of $k\langle Z\rangle$ is weakly finite from which our lemma follows. However, we know that the enveloping algebra of $k\langle Z\rangle$ embeds in a skew field, because it is a simple ring (the centre of $k\langle Z\rangle$ is k, as we stated in 8.2), and we have a homomorphism to a skew field $k\langle Z\rangle^{\circ}\langle Z'\rangle$ where Z' is a set in bijective correspondence with Z. So this is a weakly finite ring. Our theorem follows from 10.9.

There is another interesting result that we can prove using the techniques of this section, which gives us an invariant of skew field coproducts of f.d. division algebras over k. We consider the universal bimodule of derivations of such a skew field. After we have shown that the enveloping algebra of a skew field coproduct of f.d. division algebras over k is a weakly finite hereditary ring with a unique rank function, it follows that the rank of the universal bimodule of derivations, which may be considered as a f.g. projective module over the enveloping algebra, is an invariant of the skew field. Since this number may be computed without any difficulty from any coproduct representation, this is quite useful.

<u>Theorem 10.12</u> Let $\{D_i : i = 1 \text{ to } n\}$ be a finite collection of f.d. division algebras over k; then the enveloping algebra of $\mathop{\circ}_k D_i$ is a weakly finite hereditary ring with a unique rank function taking values in $\frac{1}{m}\mathbb{Z}$ for some natural number m.

Proof: By 10.10, $\mathop{\circ}_k D_i$ embeds in $M_m(k\langle Z\rangle)$ for some set Z; by the argument of 10.11, its enveloping algebra must be weakly finite since it embeds in the enveloping algebra of $M_m(k\langle Z\rangle)$.

The enveloping algebra of $\mathop{\circ}_k D_i$ is a universal localisation of $(\mathop{\circ}_k D_i)^{o} \otimes_k (\mathop{\sqcup}_k D_i)$, which is isomorphic to $\mathop{\sqcup}_{(\circ_k D_i)^{o}} ((\mathop{\circ}_k D_i)^{o} \otimes_k D_i)$.

Each ring $(\mathop{\circ}_k D_i)^{o} \otimes_k D_i$ is simple artinian, because, the centre of $\mathop{\circ}_k D_i$ is k or $k(t)$ for some transcendental t by 8.3; so our enveloping algebra is a weakly finite universal localisation of a ring with a unique rank function. It must be a localisation at some set of maps full with respect to the rank function, and so, it has a unique rank function itself by the fact that all f.g. projectives are stably induced from the ring of which it is a universal localisation by 5.10.

Once we have this theorem, we know that the universal bimodule of derivations of $\mathop{\circ}_k D_i$ over k is a projective bimodule, since it is a sub-bimodule of a free bimodule, because of the exact sequence:

$$0 \to \Omega_k(\mathop{\circ}_k D_i) \to \mathop{\circ}_k D_i \otimes_k \mathop{\circ}_k D_i \to \mathop{\circ}_k D_i \to 0.$$

It is f.g. projective from the fact that $\Omega_k(\mathop{\circ}_k D_i) \cong \oplus\, {}^{\otimes}\Omega_k(D_i)^{\otimes}$

by 10.4 and 10.6. Its rank is an invariant since there is a unique rank function.

<u>Theorem 10.13</u> The rank of $\Omega_k(\circ_k D_i)$ is $\sum_i \frac{m_i-1}{m_i}$ where $m_i = [D_i:k]$.

Proof: We have the formula $\Omega_k(\circ_k D_i) \cong \oplus\, {}^{\otimes}\Omega_k(D_i)^{\otimes}$.

From the exact sequence $0 \to \Omega_k(D_i) \to D_i \otimes_k D_i \to D_i \to 0$, we find the exact sequence $0 \to (\circ_k D_i)\otimes_{D_i} \Omega_k(D_i) \to (\circ_k D_i)\otimes_k D_i \to \circ_k D_i \to 0$ by tensoring on the left by $(\circ_k D_i)\otimes_{D_i}$. As a sequence of $\circ_k D_i$, D_i bimodules, it is a split exact sequence of f.g. projective bimodules since $(\circ_k D_i)^{\circ}\otimes_k D_i$ is simple artinian. The middle module is free of rank 1, the right module has rank $1/m_i$; so $(\circ_k D_i)\otimes_{D_i}\Omega_k(D_i)$ has rank $1 - 1/m_i$ as a $\circ_k D_i$, D_i bimodule. This implies that the rank of the bimodule ${}^{\otimes}\Omega_k(D_i)^{\otimes}$ must be $1 - 1/m_i$, and summing gives us the result we want.

We shall discuss in a later chapter the question of distinguishing more fully between skew field coproducts with f.d. factors, as well as providing examples of isomorphisms between certain of them that have the appearance of being quite different.

As a last result, we are able to show that certain skew subfields of $E\langle N\rangle$ have the form they should have by the methods of this chapter.

<u>Theorem 10.14</u> Let $M \subset N$ be an extension of bimodules over the skew field E; then the skew subfield of $E\langle N\rangle$ generated by E and M is isomorphic in the natural way to $E\langle M\rangle$.

Proof: $E<M>$ and $E<N>$ are both firs and their universal skew fields of fractions are $E\langle M\rangle$ and $E\langle N\rangle$. By 10.5 and 10.6, $\Omega_E(E\langle M\rangle) \cong {}^{\otimes}\Omega_E(E<M>)^{\otimes} \cong {}^{\otimes}M^{\otimes}$, and a similar statement holds for $\Omega_E(E\langle N\rangle)$; moreover, the universal derivation takes the form of the identity map from M to M in this representation.

Let K be the skew subfield of $E\langle N\rangle$ generated by M over E; we wish to find $\Omega_E(K)$, and we have a homomorphism from it to the K,K subbimodule of $\Omega_E(E\langle N\rangle)$ generated by δM, since M generates K over E. However, this must be simply ${}^{\otimes}M^{\otimes}$; so the map ${}^{\otimes}\Omega_E(E<M>)^{\otimes} \to \Omega_E(K)$ is an isomorphism, and, by 10.8, K must be $E\langle M\rangle$.

11 COMMUTATIVE SUBFIELDS AND CENTRALISERS IN SKEW FIELD COPRODUCTS

In this chapter, we shall find out what we can about the commutative subfields of skew field coproducts, and centralisers in the matrix rings over the free skew field; in the first problem, the transcendence degree of commutative subfields of the skew field coproduct is shown to be essentially determined by that of the factor skew fields; in the second problem we find that skew subfields of $M_n(k\langle X\rangle)$ that have centres transcendental over k are f.d. over their centre and in fact this dimension must divide n^2. In addition, we shall prove an odd result on skew subfields of the free skew field; we shall see that every 2 generator skew subfield of the free skew field is either free on those 2 generators or else it is commutative; there is an analogous result to this known for the free algebra.

The basic result we need to prove our theorems is a characterisation of the transcendence degree of commutative subfields of matrix rings over a given skew field due to Resco (80) which we shall not prove.

<u>Theorem 11.1</u> The maximal transcendence degree of a commutative subfield of $M_n(E)$ for varying n is the first integer m such that the global dimension of $E\otimes_k k(x_1 \ldots x_{m+1})$ is m, where $\{x_i\}$ is a set of independent commuting variables. If there is no such integer, the maximal possible transcendence degree is infinite.

At present, it is unknown whether there can be no commutative subfield of transcendence degree m inside $M_n(E)$ when no such field lies in E; however, there is a candidate for a counter-example. We consider $M_n(F)$ which is isomorphic to $M_n(k) \circ_k k(x_1 \ldots x_m)$. It is unclear whether $k(x_1 \ldots x_m)$ can be embedded in F.

We shall apply 11.1 to the study of commutative subfields of simple artinian coproducts; the author's original proof of the main theorem in this direction was missing a step, which was supplied by Dicks who pointed out the next theorem.

Theorem 11.2 Let R be a weakly semihereditary k-algebra, and let S be a simple artinian universal localisation of R; then the global dimension of $S\otimes_k A$ is bounded by the global dimension of $R\otimes_k A$ for any k-algebra A.

Proof: If the global dimension of $R\otimes_k A$ is ∞ or 0, the result is trivial. So, we assume that the global dimension of $R\otimes_k A$ is $n \geq 1$. Let M be some $S\otimes_k A$ module, and let $0 \to P_{n+1} \to P_n \to \ldots . \to P_1 \to M \to 0$ be a resolution of M as an $R\otimes_k A$ module, where each P_i is projective. Tensoring over R with S gives us the sequence $0 \to P_{n+1}\otimes_R S \to P_n\otimes_R S \to \ldots . \to P_1\otimes_R S \to M\otimes_R S \to 0$ which we shall show is an exact sequence. For $\mathrm{Tor}_1^R(M,S) = 0$ for $i \geq 2$, since R is weakly semi-hereditary, and so, of weak dimension 1; whilst $\mathrm{Tor}_1^R(S,S) = 0$ since S is a universal localisation of R and so, $\mathrm{Tor}_1^R(M,S) = 0$, because M is an S-module.

$$M\otimes_R S \cong M \text{ as } S\otimes_k A \text{ module, and } P_i\otimes_R S \cong P_i\otimes_{R\otimes_k A}(S\otimes_k A) \text{ which is}$$

a projective $S\otimes_k A$ module. So, M has global dimension at most n, as we wished to show.

This applies to the study of commutative subfields of simple artinian coproducts in the following way.

Theorem 11.3 Let S be simple artinian, and let S_1 and S_2 be simple artinian S-rings. Let k be a common central subfield. Suppose that the global dimension of $S\otimes_k E = n$, and the global dimension of $S_i\otimes_k E$ is n_i, then $n_i \geq n$. If $n_i = n$ for $i = 1,2$, then the global dimension of $(S_1 \underset{S}{\circ} S_2)\otimes_k E$ is n or $n+1$. If $n_i > n$ for $i = 1$ or 2, then the global dimension of $(S_1 \underset{k}{\circ} S_2)\otimes_k E$ is the maximum of n_1 and n_2.

Proof: Let $A \subset B$ be a pair of rings such that B is left free and also right free over A; then by 9.39 of Rotman (79), we know that the global dimension of A is less than or equal to the global dimension of B provided that A has finite global dimension. In particular, this applies to $S\otimes_k E \subseteq S'\otimes_k E$, where S' is any S-ring. It also applies to show that the global dimension of $(S_1 \underset{k}{o} S_2)\otimes_k E$ is at least the maximum of n_1 and n_2.

However, by the last theorem, the global dimension of $(S_1 \underset{k}{o} S_2)\otimes_k E$ is bounded by the global dimension of $(S_1 \underset{S}{\sqcup} S_2)\otimes_k E$. $(S_1 \underset{S}{\sqcup} S_2)\otimes_k E \cong (S_1\otimes_k E) \underset{S\otimes_k E}{\sqcup} (S_2\otimes_k E)$; $S_i\otimes_k E$ is free on either side over the subring $S\otimes_k E$, so, by a theorem of Dicks (77), the global dimension of $(S_1 \underset{S}{\sqcup} S_2)\otimes_k E$ is bounded by the maximum of n_1 and n_2 if one of these is larger than n, and if $n_1 = n = n_2$, then it is bounded by $n+1$. The theorem follows.

If we take E to be $k(x_1 \ldots x_m)$ for varying m, and apply 11.1 and 11.3, we deduce:

Theorem 11.4 Let n, n_1 and n_2 be the maximal transcendence degree over k of commutative subfields of $M_t(S)$, $M_t(S_1)$ and $M_t(S_2)$ for varying t, where all rings are k-algebras, and S_i is an S-ring. Then, if n_1 or n_2 is larger than n, the maximal transcendence degree of commutative subfields of $M_t(S_1 \underset{k}{o} S_2)$ over k for varying t is the maximum of n_1 and n_2; if $n_1 = n = n_2$ it may be n or $n+1$.

Proof: By 11.1, the maximal transcendence degree of commutative subfields of $M_t(S_1 \underset{S}{o} S_2)$ for varying t is equal to the maximal global dimension of $(S_1 \underset{S}{o} S_2)\otimes_k k(x_1 \ldots x_m)$ for varying m. The theorem follows at once from 11.3.

Both possibilities may occur for $n_1 = n = n_2$. First, consider $S = k(x_1 \ldots x_m)$ and $S_i = S \underset{k}{o} k(y_i)$; then $S_1 \circ S_2 \cong k(y_1) \underset{k}{o} k(y_2) \underset{k}{o} k(x_1 \ldots x_m)$. By 11.4, the transcendence degree of maximal commutative subfields of $M_t(S_1 \underset{S}{o} S_2)$ is at most m in this case. On the other hand, let $S = k(x_1 \ldots x_m)$, and $S_i \cong S(\alpha_i)$ where $\alpha_i^2 = x_1$. Then the centre of $S_1 \underset{S}{o} S_2$ contains S, and also the element $\alpha_1\alpha_2 + \alpha_2\alpha_1$, which is transcendental over S. So, here, the maximal transcendence degree of commutative subfields is $n+1$.

We pass from considering the commutative subfields of skew field coproducts to the study of centralisers. There have been a number of interesting results on centralisers in skew fields and rings. The most interesting of these is due to Bergman (67), who showed that the centraliser of a non-central element in the free algebra over a field is a polynomial ring in one variable. Another result of interest is due to Cohn (77') who showed that the centraliser of a non-central element in the free skew field is commutative; together with the earlier results of this chapter, we also know that its transcendence degree over k is 1. We shall prove that it must also be f.g. as a special case of results on centralisers in skew subfields of $M_n(k\langle X\rangle)$. We shall show that centralisers of elements transcendental over k in such skew fields are f.g. of p.i. degree dividing n and having transcendence degree 1 over k. We have already seen that a number of skew fields may be embedded in $M_n(k\langle X\rangle)$. In particular, we showed in 7.12 that all skew field coproducts of f.d. division algebras over k may be so embedded.

<u>Theorem 11.5</u> Let C be an arbitrary commutative field extension of k; then $C\otimes_k M_n(k\langle X\rangle)$ is an hereditary noetherian prime ring.

Proof: By 10.8, $k\langle X\rangle$ is the skew subfield of $C\langle X\rangle$ generated by k and X. Since $k\langle X\rangle$ has centre k, $C\otimes_k k\langle X\rangle$ is a simple ring and so, must embed in $C\langle X\rangle$, which shows that it is a domain. It must be an Ore domain, because C is commutative, and therefore, by representing C as a direct limit of f.g. fields $C\otimes_k k\langle X\rangle$ may be represented as a direct limit of noetherian domains.

$C\otimes_k k\langle X\rangle$ is a universal localisation of $C<X>$, and so, it must be hereditary by 4.9. However, an hereditary Ore domain must be noetherian by Robson (68). Therefore, $C\otimes_k M_n(k\langle X\rangle) \cong M_n(C\otimes_k k\langle X\rangle)$ is an hereditary noetherian prime ring as we wished to show.

This gives us another handle on commutative subfields of $M_n(k\langle X\rangle)$.

<u>Theorem 11.6</u> Any commutative subfield of $M_n(k\langle X\rangle)$ is f.g. of transcendence degree at most 1 over k.

Proof: Let C be a commutative subfield of $M_n(k\langle X\rangle)$; by 11.5, $C\otimes_k M_n(k\langle X\rangle)$ is hereditary and noetherian. So, $C\otimes_k C$ must be noetherian, since $C\otimes_k M_n(k\langle X\rangle)$ is faithfully flat over it (see Resco, Small, Wadsworth 79)). Therefore, C must be f.g. That the transcendence degree is at most 1 follows from 11.4.

This already shows that centralisers in $k\langle X\rangle$ of non-central elements are f.g. and so, f.d. over the field generated by the element they centralise. We extend this a little.

Theorem 11.7 Let E be a skew subfield of $M_n(k\langle X\rangle)$ whose centre contains an element y transcendental over k; then $[E:k(y)]$ is finite.

Proof: Let M be a maximal commutative subfield of E. $M \geq k(y)$, and so, is f.g. of transcendence degree 1 over k. Therefore, M is a f.d. algebraic extension of $k(y)$.

If C is the centre of E, $[M:C] \leq [M:k(y)]$ and is finite; therefore, if M' is the centraliser of M in E, $[E:M'] = [M:C]$ is finite. But $M' = M$, and so, $[E:k(y)] = [E:M][M:k(y)]$ is finite too.

So, skew subfields of $M_n(k\langle X\rangle)$ whose centres are transcendental over k are f.d. over their centre. It turns out that we can get a good bound on the p.i. degree of such skew fields. The result we prove yields Cohn's theorem that centralisers in the free skew field are commutative.

Theorem 11.8 Let E be a skew subfield of $M_n(k\langle X\rangle)$ of finite p.i. degree. Then the p.i. degree divides n.

Proof: Let $\bar{k}$ be the algebraic closure of k; E embeds in $M_n(k\langle X\rangle)$, and so, $E\otimes_k\bar{k}$ embeds in $M_n(k\langle X\rangle\otimes_k\bar{k} \cong M_n(\bar{k}\langle X\rangle)$.

If E has finite p.i. degree, E is f.d. over its centre, which is f.g. over k by 12.6, and so, $E\otimes_k\bar{k}$ must be artinian.

If $E\otimes_k\bar{k}/\mathrm{rad}(E\otimes_k\bar{k}) \cong \times_i S_i$, where each S_i is simple artinian, each S_i is a central extension of E, and so, has the same p.i. degree. Further, the transcendence degree of the centre of each S_i is at most 1, so, by Tsen's theorem, S_i is isomorphic to $M_m(C_i)$ where m is the p.i. degree of E, and C_i is the centre of S_i. Therefore, $E\otimes_k\bar{k}/\mathrm{rad}(E\otimes_k\bar{k}) \cong M_m(\times_i C_i)$, and since matrix units lift modulo a nilpotent ideal, $E\otimes_k\bar{k} \cong M_m(A)$ for some artinian ring A. Therefore, we have a unit-preserving embedding of $M_m(\bar{k})$ in $M_n(\bar{k}\langle X\rangle)$, where m is the p.i. degree of E, which shows that m divides n.

We summarise the last few results in the following form.

Theorem 11.9 Let E be a skew subfield of $M_n(k\langle X\rangle)$ whose centre is transcendental over k; then E is f.d. over its centre which is f.g. of transcendence degree 1 over k, and the p.i. degree of E divides n.

We come to the last theorem of this chapter which is on 2 generator skew subfields of the free skew field. At present, it is quite unclear whether all skew subfields of the free skew field must themselves be isomorphic to free skew fields on some number of generators; we have already seen that commutative subfields must be f.g. of transcendence degree 1 but the methods are clearly too weak to show that they actually must be rational. It therefore comes as a surprise that we are able to prove anything at all about 2 generator skew subfields.

Theorem 11.10 Let F be a 2 generator skew subfield of $k\langle X\rangle$ over k; then either F is commutative or else it is free on the 2 generators.

Proof: $F^o\otimes_k k\langle X\rangle$ is a weakly finite hereditary ring such that all projectives are stably free of unique rank, because it is a universal localisation of $F^o<X>$ which embeds in $F^o\langle X\rangle$.

We have an exact sequence of F, $k\langle X\rangle$ bimodules:

$$0 \to \Omega_k(F)^{\otimes} \to F\otimes_k k\langle X\rangle \to k\langle X\rangle \to 0$$

where the left action of F on $k\langle X\rangle$ is given by the embedding of F in $k\langle X\rangle$. Since F is generated by elements s and t, $\Omega_k(F)$ is generated by elements δs, δt. By theorem 10.7, either F is freely generated as a skew field by s and t or else there is a non-trivial relation between δs and δt; in the latter case, $\Omega_k(F)^{\otimes}$ is a non-zero 2 generator projective bimodule with some non-trivial relation between the 2 generators; so it has rank 1 considered as a projective $F^o\otimes_k k\langle X\rangle$ module, and $k\langle X\rangle$ considered as an $F^o\otimes_k k\langle X\rangle$ module via the left action of F on $k\langle X\rangle$ described above is a simple module that is torsion with respect to the rank function.

We have another torsion $F^o\otimes_k k\langle X\rangle$ module, $F\otimes_{k(s)} k\langle X\rangle$ and this maps onto our simple module via the multiplication map.

The category of torsion modules is an abelian finite length category by 1.22; therefore, as a torsion module, $F\otimes_{k(s)} k\langle X\rangle$ has a unique largest semi-simple quotient torsion module and the endomorphism ring of

$F \otimes_{k(s)} k\langle X \rangle$ must act on this semisimple module. This semisimple torsion module maps non-trivially onto our simple module ${}_F k\langle X \rangle_{k\langle X \rangle}$; it has a unique largest direct summand of the form $({}_F k\langle X \rangle_{k\langle X \rangle})^n$, and the endomorphism ring of $F \otimes_{k(s)} k\langle X \rangle$ must also act on this module. So, we have a map from $k(s)$ which lies in the endomorphism ring of $F \otimes_{k(s)} k\langle X \rangle$ to the endomorphism ring of $({}_F k\langle X \rangle_{k\langle X \rangle})^n$ which is $M_n(C)$ where C is the centraliser of F in $k\langle X \rangle$; consequently, $[C:k] = \infty$ because s must be transcendental over k, and since centralisers of elements of $k\langle X \rangle$ that are not in k are commutative, F, which centralises C, must be commutative.

12 CHARACTERISING UNIVERSAL LOCALISATIONS AT A RANK FUNCTION

Simple artinian universal localisations

In chapter 10, we saw that if R is a right hereditary k-algebra, we can characterise those epic skew fields over R that are universal localisations by the property that the associated map on the universal bimodule of derivations ${}^{\otimes}\Omega_K(R)^{\otimes} \to \Omega_K(F)$ is an isomorphism. As we shall see, this is equivalent to the property that $\mathrm{Tor}_1^R(F,F) = 0$, which is a condition that we have already discussed in chapter 4; this is precisely the property that we need to generalise to a characterisation of epic simple artinian rings that are universal localisations. This condition fits nicely into the theory of f.d. hereditary algebras and allows us to characterise those epimorphisms from f.d. hereditary algebras to f.d. simple algebras that arise as universal localisations.

In order to prove these results, we have to discover a lot of information about the module structure of epic simple artinian rings that are universal localisations at a rank function on a hereditary ring, R; we are able to turn this information into results about epic R-subrings of such simple artinian rings; they must all be universal localisations of R. Such a result was already known in the Noetherian case where, however, it was stated in terms of Silver localisation. One consequence of this result on intermediate rings is that an epic endomorphism of the free algebra on n generators over a field k must be an isomorphism.

It may well be true that if R is a right hereditary ring, and the map $R \to R'$ is an epimorphism of rings such that $\mathrm{Tor}_1^R(R',R') = 0$ then R' is forced to be a universal localisation of R. The author has no way of attacking this question, however, and the only results are those in this chapter apart from the result that if $I = I^2$ for an ideal I in a right hereditary ring, R, then I is a trace ideal which shows that R/I is a universal localisation of R as we saw in chapter 4. Such a result cannot be true for all semihereditary rings as one sees by looking at a local Bezout domain, L, whose maximal ideal, M, is idempotent; $\mathrm{Tor}_1^L(L/M,L/M) = 0$, since $M^2 = M$.

<u>Lemma 12.1</u> Let $\phi:R \to S$ be an epimorphism between K-rings, where K is a semisimple artinian ring; then $\Omega_K(S) \cong {}^{\otimes}\Omega_K(R)^{\otimes}$, if and only if $\mathrm{Tor}_1^R(S,S) = 0$.

Proof: By 7.1, $0 \to \Omega_K(R) \to R\otimes_K R \to R \to 0$ is an exact sequence split as a sequence of left R-modules, so the sequence $0 \to S\otimes_R\Omega_K(R) \to S\otimes_K R \to S \to 0$ is also exact, and since K is semisimple artinian, $S\otimes_K R$ is a projective R-module. Therefore, we have an exact sequence:

$0 \to \mathrm{Tor}_1^R(S,S) \to {}^{\otimes}\Omega_K(R)^{\otimes} \to S\otimes_K S \to S \to 0.$

We extract the exact sequence:

$0 \to \mathrm{Tor}_1^R(S,S) \to {}^{\otimes}\Omega_K(R)^{\otimes} \to \Omega_K(S) \to 0$, from which our lemma follows.

Before setting about the main proofs, we isolate a useful lemma.

<u>Lemma 12.2</u> Let R be a ring of weak dimension 1 (in particular, a right hereditary ring), let M and N be a right and left module over R, respectively, with submodules M' and N'; then if $\mathrm{Tor}_1^R(M,N) = 0$, $\mathrm{Tor}_1^R(M',N') = 0$.

Proof: Apply the long exact sequence of $\mathrm{Tor}_i^R(\ ,N)$ to the sequence $0 \to M' \to M \to M/M' \to 0$, which shows that $\mathrm{Tor}_1^R(M',N) = 0$, and then apply the long exact sequence of $\mathrm{Tor}_i^R(M',\ )$ to the sequence $0 \to N' \to N \to N/N' \to 0$ to complete the proof.

We can begin on the first result.

<u>Theorem 12.3</u> Let R be a right hereditary ring and let $\phi:R \to S$ be an epimorphism from R to a simple artinian ring S; then S is a universal localisation of R if and only if $\mathrm{Tor}_1^R(S,S) = 0$.

Proof: Universal localisations of a ring R at some set of maps Σ between f.g. projectives always satisfy the condition $\mathrm{Tor}_1^R(R_\Sigma,R_\Sigma) = 0$ by Bergman and Dicks (78) and 4.7. The strategy for proving the converse is to find the structure of S as a right and left R module.

First of all, we can divide out by the trace ideal, T,, of all f.g. projectives of rank 0 with respect to ρ, the rank function induced on R by the map to S; this is a universal localisation of R at suitable maps all of which become invertible over S. By 4.9, a universal localisation of a right hereditary ring is right hereditary, so R/T is right hereditary; by 1.7 and 1.8, all f.g. projectives over R/T are induced from R, and the rank function ρ on R induces a faithful rank function on R/T. Since T lies in the kernel of $R \to S$ and $\mathrm{Tor}_1^R(S,S) = 0$, $\mathrm{Tor}_1^{R/T}(S,S) = 0$ too. Therefore, if we may prove our theorem on the assumption that ρ is a faithful rank function, it follows in general.

Let M be a right R submodule of S; then, we see from 12.2 that $\mathrm{Tor}_1^R(M,S) = 0$. Let $0 \to P \overset{\alpha}{\to} Q \to M \to 0$ be a presentation of M, where Q is f.g. projective, and P must be projective, and therefore, a direct sum of f.g. projectives by 1.2. Then, since $\mathrm{Tor}_1^R(M,S) = 0$, the sequence below is exact:

$$0 \to P\otimes_R S \overset{\alpha\otimes S}{\to} Q\otimes_R S \to M\otimes_R S \to 0 .$$

Since ρ is a faithful rank function, and $M\otimes_R S \neq 0$, P must be finitely generated, and $\rho(P) < \rho(Q)$. Also, α must be a left full map, for, if it were not, $\alpha\otimes_R S$ factors through a module of smaller rank, and cannot be injective.

We define the presentation rank of a f.p. module by $p.\rho(M) = \rho(Q) - \rho(P)$, where $0 \to P \to Q \to M \to 0$ is a presentation of M; it is well-defined by Schanuel's lemma. Our aim is to show that S as a right module is a directed union of f.p. modules of left full presentation of presentation rank 1, having no submodules of presentation rank 0. There is an analogous result on the left.

In order to carry this out, we need to see how left full maps with respect to ρ behave under $\otimes_R S$. We wish to show that they become injective; this is a little harder to show immediately than that right full maps become surjective, which is our next step.

Let $\alpha: P \to Q$ be a right full map between f.g. right projectives over R; then all f.g. submodules of Q containing the image of P have rank at least that of Q, and so, by our first step, cannot occur as the kernel of a map from Q to S. Since our first step showed that all such kernels are finitely generated, we deduce that $\mathrm{Hom}(\mathrm{coker}\alpha, S) = 0$; since all modules over S are projective, this shows that $\mathrm{coker}\alpha\otimes_R S = 0$, and

$\alpha \otimes_R S$ must be surjective.

Since all right full maps between f.g. right projectives become split surjective over S, all left full maps between left f.g. projectives become split injective over S by duality. If we had assumed two-sided hereditary, we would know that left full maps between f.g. right projectives become split injective, and we would proceed as we do in the latter half of the proof; as it is, we are able to deduce these properties in the right hereditary case with a little more work.

So, let M be a f.g. left R submodule of S, then by 12.2, $\mathrm{Tor}_1^R(S,M) = 0$. Let $0 \to F \to Q \to M \to 0$ be a presentation of M, where Q is f.g. projective, and, since R is left semihereditary by 1.8, all f.g. submodules of F are projective. Our intention is to show that F is a directed union of f.g. projectives where all inclusions in the system are left full, and the rank of each module in the system is less than that of Q.

Let $\{P_{1i}:i \in I_1\}$ be the set of f.g. submodules of F of minimal possible rank q_1; if there are f.g. submodules of F that do not lie in such a P_{1i}, we consider the set of f.g. submodules $\{P_{2i}:i \in I_2\}$ where $P_{2i} \not\subseteq P_{1j}$ for any i,j and the rank of each P_{2i} is q_2, the minimal possible. In general, at the nth stage, either all f.g. submodules are inside some P_{ki} for $k < n$, or else, we form the set of f.g. submodules $\{P_{ni} : i \in I_n\}$ such that $P_{ni} \not\subseteq P_{kj}$ for $k < n$, and the rank of P_{ni} is q_n, the minimal possible. We see that in the limit, every f.g. submodule of F must lie in some P_{ni}, since the ranks q_n form an ascending sequence of numbers in $\frac{1}{m}\mathbb{N}$, and so, are eventually bigger than the rank of any f.g. module. So F is the directed union of these modules; moreover, if $P_{kl} \subseteq P_{ij}$, the inclusion is a left full map, since there are no intermediate modules of smaller rank.

Consequently, $S \otimes_R F$ is the directed union of the system of f.g. modules $\{S \otimes_R P_{ki}:k \in \mathbb{N},\ i \in I_k\}$ and all the maps in this system are injective, because we have shown that left full maps between f.g. left projectives become injective. As noted above, $\mathrm{Tor}_1^R(S,M) = 0$, so the sequence below is exact:

$$0 \to S \otimes_R F \to S \otimes_R Q \to S \otimes_R M \to 0 .$$

Therefore, $S \otimes_R F$ must be finitely generated and of rank less than that of Q, since $S \otimes_R M$ does not equal 0. Consequently, the rank of each P_{ki} must be less than the rank of Q, since $S\otimes_R P_{ki}$ embeds in $S\otimes_R F$, our directed system must have stopped at some finite stage, and it represents F as a directed system of f.g. submodules where all maps are left full, and the modules all have rank less than that of Q, as we wished.

Let $\alpha:P \to Q$ be a right full map between left f.g. projectives; then all f.g. submodules containing the image of α have rank at least that of Q, so that the image of α cannot lie in the kernel of a map from Q to S. Therefore, $\text{Hom}_R(\text{coker}\alpha, S) = 0$, and $S\otimes_R \text{coker}\alpha = 0$. Right full maps between left f.g. projectives become split surjective, and so, by duality, left full maps between left f.g. projectives become split injective, as we wished to show originally.

Let $M \subseteq N$ be a pair of f.p. right modules of left full presentation such that the presentation rank of any module M_1 between M and N is at least $p.\rho(M)$; then we show next that $M\otimes_R S \to N\otimes_R S$ is an embedding. For consider the commutative diagram with exact rows:

$$\begin{array}{ccccccccc} 0 & \to & P & \xrightarrow{\alpha} & Q_1 & \to & M & \to & 0 \\ & & \| & & \downarrow & & \cap & & \\ 0 & \to & P & \xrightarrow{\rho} & Q_2 & \to & N & \to & 0, \end{array}$$

obtained by pullback from a presentation of N, where α,β are left full maps. Then, any Q such that $Q_1 \subseteq Q \subseteq Q_2$ satisfies $\rho(Q) \geq \rho(Q_1)$ since the presentation rank of the image of Q in N is at least that of M by assumption. So, $Q_1 \to Q_2$ is left full. Tensoring our diagram with S gives us a commutative diagram with exact rows:

$$\begin{array}{ccccccccc} 0 & \to & P\otimes_R S & \to & Q_1\otimes_R S & \to & M\otimes_R S & \to & 0 \\ & & \| & & \cap & & \downarrow & & \\ 0 & \to & P\otimes_R S & \to & Q_2\otimes_R S & \to & N\otimes_R S & \to & 0, \end{array}$$

since α and β are left full. It follows that the map $M\otimes_R S \to N\otimes_R S$ must be an embedding. We use this to examine the structure of S as right R module in a similar way to our method of studying a left submodule of a free left module earlier on.

First of all, we note that every right R submodule of S containing R has presentation rank at least 1; for on tensoring $R \subseteq M \subseteq S$ with S, we obtain $S = R\otimes_R S \to M\otimes_R S \to S\otimes_R S = S$, where the composite map is the identity, so that the rank of $M\otimes_R S$ is at least 1; but we have seen that $\text{Tor}_1^R(M,S) = 0$, so that $p.\rho(M) = \rho_S(M\otimes_R S) \geq 1$.

We consider the set of all f.g. right R submodules M_{1i} of S such that M_{1i} contains R, and $p.\rho(M_{1i})$ is minimal, which implies that it is 1. Our intention is to show that S is the directed union of these modules. If it is not, let $\{M_{2i} : i \in I_2\}$ be the sets of f.g. submodules of S that contain R, do not lie in any M_{1i} and have minimal presentation rank, q_2, subject to these conditions. In general, our nth step is to take the set of f.g. submodules of S that contain R, do not lie in M_{ki} for $k < n$, and have minimal rank subject to these conditions. Every f.g. submodule of S must lie in some M_{ki} since we have seen that all f.g. submodules have a presentation rank; consequently, S is the directed union of the directed system of submodules $\{M_{ki}\}$. If $M_{kj} \subseteq M_{1i}$, there can be no f.g. submodule of M_{1i} such that its presentation rank is less than that of M_{kj}; so, we have shown above that the map obtained by tensoring with S, $M_{kj} \otimes_R S \to M_{1i} \otimes_R S$ must be an embedding. This shows that S which as S module is isomorphic to $S \otimes_R S$ is the directed union of the submodules $M_{ki} \otimes_R S$; in particular, each of the ranks $\rho_S(M_{ki} \otimes_R S) = 1$. Since this is just the presentation rank of M_{ki} $(\mathrm{Tor}_1^R(M_{ki}, S) = 0)$, our process must have stopped at the first step; that is, S is the directed union of the f.g. right R submodules $\{M_{1i}\}$.

From here, it is not too hard to see that S must he a universal localisation of R. We know that all full maps between f.g. right projectives become invertible, since they are both left and right full. So we have a map from the universal localisation of R at the rank function, $R\rho$, to S, which we shall show is surjective.

Let $s \in S$, and let M be a right R submodule of presentation rank 1 of S containing both R and s; we consider the commutative diagram with exact rows:

$$\begin{array}{ccccccccc}
0 & \to & P & \to & P \oplus R & \to & R & \to & 0 \\
 & & \| & & \downarrow & & \cap\!| & & \\
0 & \to & P & \to & Q & \to & M & \to & 0
\end{array} ,$$

obtained by pullback along $R \subseteq M$ from a presentation of M. We know that $P \oplus R$ and Q have the same rank, and since all f.g. modules between R and M have presentation rank at least 1, the middle column is a full map. So, over R_ρ it is invertible, and its inverse induces a map from M to R_ρ whose composite with the homomorphism from R_ρ to S induces the embedding of M in S we began with. So the map from R_ρ to S is surjective.

We know that R_ρ is a perfect ring by 5.3, so the surjective maps from R_ρ to simple artinian rings arise as $R_\rho \to R_\rho/N \to \bar{e}R_\rho/N$, where N is the nil radical, and $\bar{e}$ is a central idempotent in the semisimple artinian ring R_ρ/N. If e is an idempotent of R_ρ whose image in R_ρ/N is $\bar{e}$, it is clear that $\bar{e}R_\rho/N \cong eR_\rho e$ is the universal localisation of R_ρ at the map e, so S is a universal localisation as we wished to show, because all f.g. projectives over R_ρ are stably induced from R by 5.3, and so, the iterated universal localisation theorem, 4.6, applies.

Of course, after the event, we know that S is the universal localisation of R at the rank function, so that R_ρ is a rather better ring than we knew during the proof.

It is possible to give a description of the right R module structure of S in the situation of theorem 12.3 essentially by abstracting the relevant information from the proof. First, we assume the rank function is faithful by passing to the quotient by the trace ideal of the f.g. projectives of rank 0 if necessary. Next, we see that S is the directed union of the f.g. submodule of S that contain R and have presentation rank 1. If M is such a submodule, all modules between R and M have presentation rank at least 1. Conversely, if $\alpha:R \to M$ is an embedding of R in a f.g. module of left full presentation having presentation rank 1, and all f.g. modules between R and M have presentation rank at least 1, we form the commutative diagram with exact rows obtained by pullback from a presentation of M:

$$\begin{array}{ccccccccc} 0 & \to & P & \to & P \oplus R & \to & R & \to & 0 \\ & & \downarrow & & \downarrow & & \cap & & \\ 0 & \to & P & \to & Q & \to & M & \to & 0 . \end{array}$$

It is clear that the embedding of $P \oplus R$ in Q is full, and so, it is invertible over S, which leads to a homomorphism from M to S extending the map from R to S. The kernel of this map is the unique maximal submodule of M of presentation rank 0. So we wish to see how all the maps $\alpha:R \to M$ from R to f.g. submodules of presentation rank 1, such that all submodules have presentation rank not 0, and those that contain R have presentation rank at least 1, having left full presentations, fit together to form S. If we have two such maps $\alpha_i:R \to M_i$, the pushout map gives us a map $\alpha:R \to M$ to a module of the right form apart from the problem that it may have a submodule of presentation rank 0, so we pass from M to the quotient by the unique maximal such submodule of M, $\bar{M}$; this gives us a commutative diagram:

$$\begin{array}{ccc} R & \to & M_1 \\ \downarrow & & \downarrow \\ M_2 & \to & \bar{M} \end{array}$$

so that we may form the directed system of all such maps. The direct limit is S.

We may combine the last theorem with 12.1.

<u>Theorem 12.4</u> Let R be a right hereditary K-ring, where K is semisimple artinian; let $R \to S$ be a map from R to a simple artinian ring S; then S is the universal localisation of R at the rank function induced by the map if and only if $\Omega_K(S) \cong {}^{\otimes}\Omega_K(R)^{\otimes}$.

Proof: We simply combine the equivalences given by 12.1 and 12.3.

In chapter 1, we defined the transpose of a f.g. module of homological dimension 1 such that the dual of the module is trivial by $\mathrm{Tr}M = \mathrm{Ext}^1_R(M,R)$. This defines a duality between the categories of such modules on the left and on the right as we showed in 1.19. A <u>pre-projective module</u>, M, is an indecomposable module such that $(D\mathrm{Tr})^n M$ is projective for some non-negative integer n; a <u>pre-injective module</u>, M, is an indecomposable module such that $(\mathrm{Tr}D)^n M$ is injective for some n. It is clear that these are dual notions with respect to the duality D. Ringel also defines a module M over a f.d. hereditary algebra to be a <u>brick</u> if $\mathrm{End}_R(M)$ is a f.d. division algebra and $\mathrm{Ext}^1_R(M,M) = 0$; pre-projective and pre-injective modules are examples of bricks.

<u>Theorem 12.5</u> Let R be a f.d. hereditary k-algebra and let M be a f.g. indecomposable module over R with endomorphism ring D a f.d. division algebra such that $[M:D] = m$; then the associated map from R to $M_m(D)$ is a universal localisation at the rank function it induces on R if and only if M is a brick; in particular, for all pre-projectives or pre-injectives, the associated homomorphism from R to a f.d. simple artinian ring is always an epimorphism.

Proof: All $M_m(K)$ modules considered as R modules are direct sums of copies of M; for M, $\mathrm{Ext}^1_R(M,M) = 0 = \mathrm{Ext}^1_{M_m(k)}(M,M)$ and so,

$\mathrm{Ext}^1_{M_n(k)} = \mathrm{Ext}^1_R$ for all $M_m(k)$ modules; by theorem 4.8,

$\mathrm{Tor}^R_1(M_m(k),M_m(k)) = 0$, and so by theorem 8.3, $M_m(k)$ is a universal localisation of R.

We shall use the last result in the next chapter to construct some interesting isomorphisms between simple artinian coproducts, but, for the present, we shall investigate the subring structure of simple artinian universal localisations of right hereditary rings more closely.

Epic subrings of simple artinian universal localisations of hereditary rings

In this section, we shall show that all epic subrings must themselves be universal localisations. The principal application of this result is to show that all epic endomorphisms of the free k-algebra on n generators must be isomorphisms; this was shown for $n = 2$ by Dicks and Lewin (82).

Theorem 12.6 Let R be a right hereditary ring with a rank function ρ such that R_ρ is a simple artinian ring; let T be an epic R subring of R_ρ; then T is the universal localisation of R at those maps between f.g. projectives over R that become invertible over T.

Proof: The method of proof is entirely similar to that of 12.3 except for the need in one or two places for closer attention. We may assume as we do in 12.3, that ρ is a faithful rank function.

We have already seen in the course of 12.3 that all f.g. submodules of R_ρ are f.p. of left full presentation and if M is such a module, $\mathrm{Tor}_1^R(M,R_\rho) = 0$. Further, R_ρ is the directed union of f.g. R submodules of presentation rank 1; we shall show that T is also a directed union of f.g. R submodules of presentation rank 1, from which it will follow in a similar way to the proof of 12.3, that T is a universal localisation of R.

Let $R \subseteq M \subseteq T$, where M has presentation rank 1; then, we find that the inclusion $M \subseteq R_\rho$ becomes an isomorphism $M \otimes_R R_\rho \to R_\rho \otimes_R R_\rho \cong R$, under tensoring with R_ρ over R; for $M \otimes_R R_\rho$ is isomorphic to R_ρ since M has presentation rank 1, and $\mathrm{Tor}_1^R(M,R_\rho) = 0$, and the image of $M \otimes_R R_\rho$ in R_ρ is $MR_\rho = R_\rho$, since $R \subseteq M$. Consequently, $\mathrm{Tor}_1^R(R_\rho/M,R_\rho) = 0$, as we see by considering the long exact sequence of $\mathrm{Tor}_1^R(\ ,R_\rho)$ associated to the short exact sequence: $0 \to M \to R_\rho \to R_\rho/M \to 0$; we find the exact sequence, $0 = \mathrm{Tor}_1^R(R_\rho,R_\rho) \to \mathrm{Tor}_1^R(R_\rho/M,R_\rho) \to M \otimes_R R_\rho \xrightarrow{\sim} R_\rho \otimes_R R_\rho$, which demonstrates that $\mathrm{Tor}_1^R(R_\rho/M,R_\rho) = 0$. By lemma 12.2, we see that

$\mathrm{Tor}_1^R(T/M,T) = 0$, which allows us to show that $M\otimes_R T \to T\otimes_R T \to T$ is an isomorphism; certainly, $M\otimes_R T \to T\otimes_R T \to T$ is an injective map because $\mathrm{Tor}_1^R(T/M,T) = 0$, and T is an epic R ring; the image is just MT which contains $RT = T$, since $R \subseteq M$.

We consider the set of f.g. R submodules of T $\{M_{1i} : i \in I_1\}$ such that $R \subseteq M_{1i}$ and the presentation rank of each M_{1i} is 1; we wish to show that T is the directed union of these modules.

If it is not, let $\{M_{2i} : i \in I_2\}$ be those f.g. R submodules of T that contain R, do not lie in any M_{1i}, and have minimal presentation rank, q_2, subject to these conditions; we saw in the course of the proof of 12.3, that q_2 is larger than 1. In general, at the nth stage, we consider the set of f.g. R submodules of T $\{M_{ni} : i \in I_n\}$ that contain R, do not lie in any M_{ki} for $k < n$, and have minimal presentation rank q_n, subject to these conditions. It is clear that T is the directed union of the complete set of M_{ki} over all k. We know that if $M_{k_1j} \subseteq M_{k_2i}$, then $k_1 < k_2$ and any module between the two of them has rank at least equal to q_{k_1}; consequently, as we saw in the course of the proof of 12.3, $M_{k_1j}\otimes_R R_\rho \to M_{k_2i}\otimes_R R_\rho$ must be an embedding of a module of rank q_{k_1} in one of rank q_{k_2}. Therefore, $T\otimes_R R_\rho$ $T\otimes_R T\otimes_T R_\rho \cong T\otimes_T R_\rho \cong R_\rho$ (since T is an epic R ring) must be the directed union of the system of modules $M_{ki}\otimes_R R_\rho$, and so, each such R_ρ module has rank at most 1, which implies as we wanted, that T must be the directed union of the submodules M_{1i} that contain R and have presentation rank 1.

By the same method as we employed in the proof of 12.3, we deduce that if Σ is the collection of maps between f.g. projectives over R that become invertible over T, the map from R_Σ is surjective. All these maps in Σ must be full with respect to ρ, since they are invertible over R_ρ, so we may apply the result proved in 5.8, that the image of an intermediate localisation in the complete localisation at all full maps is still a universal localisation; in particular, T is a universal localisation.

Together with the theorems at the end of chapter 5, we see that the epic R-subrings of the simple artinian universal localisation of a hereditary ring R are precisely the universal localisations of R at a factor closed set of full maps between f.g. projectives.

It is not a great deal of effort to prove from this theorem that all epic endomorphisms of the free algebra on n generators must be isomorphisms.

<u>Theorem 12.7</u> Let $\phi: k\langle x_1 \ldots . x_n\rangle \to k\langle x_1 \ldots . x_n\rangle$ be an epic endomorphism, then it is an isomorphism.

Proof: The elements $\phi(x_i)$ generate the free skew field $k\{x_1 \ldots . x_n\}$; consequently, by 10.11, they must be free generators; therefore $k\langle x_1 \ldots . x_n\rangle$ is an epic $k\langle\phi(x_1)\ldots\phi(x_n)\rangle$ subring of its universal skew field of fractions, and, by 12.6, it must be a universal localisation of $k\langle\phi(x_1)\ldots\phi(x_n)\rangle$ at some set of maps. In order to show that it can only be the trivial localisation, we consider the induced map on the functor K_1, $K_1(\phi)$.

Since $K_1(k\langle x_1 \ldots . x_n\rangle) = k^*$, by Gersten (74), $K_1(\phi)$ must be the identity map on k^*; on the other hand, we have an exact sequence of K-groups associated to the universal localisation by 4.11:

$$K_1(R) \to K_1(R_\Sigma) \to K_0(\underline{T}) \to K_0(R) \to K_0(R_\Sigma) \enspace ,$$

where R and R_Σ are isomorphic to $k\langle x_1 \ldots . x_n\rangle$ and $\underline{T}$ is the full subcategory of modules of the form cokerα where α is in Σ; $K_0(R) \to K_0(R_\rho)$ is injective, and $K_1(\phi)$ is an isomorphism so that $K_0(\underline{T})$ must be zero; but this implies that Σ must be trivial as required because it implies that $k\langle x_1 \ldots x_n\rangle$ is equal to $k\langle\phi(x_1)\ldots .\phi(x_n)\rangle$. So ϕ is an isomorphism.

13 BIMODULE AMALGAM RINGS AND ARTIN'S PROBLEM

One of the purposes of this chapter is to construct a new class of skew fields generalising the skew field coproduct with amalgamation. They are interesting to us for several reasons. Many arise naturally as skew subfields of skew field coproducts; also, the methods that apply naturally to them apply just as well to the skew field coproduct giving us results that would not have been clear without this greater generality; further, we may use the new construction in order to show a number of interesting isomorphisms between apparently different skew field coproducts, and to study the simple artinian coproducts of the form $M_m(k) \circ_k M_n(k)$. However, the major interest in the construction is the solution it leads to for Artin's problem. Artin asked whether there are skew field extensions, $E \supset F$, such that the left and right dimension of E over F, respectively $[E:F]$ $[E:F]_r < \infty$ but $[E:F]_l \neq [E:F]_r$. We shall see that for arbitrary pairs of integers greater than 1 occur as the left and right dimension of a skew field extension. In recent work of Dowbor, Ringel and Simson (79), it was shown that the hereditary artinian rings that have only finitely many indecomposable modules correspond to Coxeter diagrams in the same way that hereditary artinian algebras (f.d. over a central subfield) correspond to Dynkin diagrams; they were unable to show however that any Coxeter diagram that is not a Dynkin diagram actually had a corresponding hereditary artinian ring, since the existence of such a hereditary artinian ring required the existence of an extension of skew fields having different but finite left and right dimension together with further conditions; at the end of the chapter, there is an example of an hereditary artinian ring of finite representation type corresponding to the Coxeter diagram $I_2(5)$.

Bimodule amalgam rings

Given a couple of skew fields E_1 and E_2, we define a <u>pointed cyclic</u> E_1, E_2 <u>bimodule</u> to be a pair (M,x) where M is a E_1, E_2 bimodule,

x is in M, and $M = E_1 x E_2$. For example, if F is a common skew subfield of E_1 and E_2, the pair $(E_1 \otimes_F E_2, 1\otimes 1)$ is a pointed cyclic bimodule on an F-centralising generator.

The importance of this idea for us lies in the observation that the ring coproduct $E_1 \underset{F}{\sqcup} E_2$ is the universal ring containing a copy of E_1 and E_2 such that the pointed cyclic bimodule $(E_1,E_2,1)$ is a quotient of $(E_1 \otimes_F E_2, 1\otimes 1)$.

If (M,x) is a pointed cyclic E_1, E_2 bimodule and the relations for the generator x are $\sum_j e_{ij} x e'_{ij} = 0$; $e_{ij} \in E_1$, $e'_{ij} \in E_2$, then the universal ring containing a copy of E_1 and E_2 such that $(E_1E_2,1)$ is a quotient of (M,x) is clearly the ring

$$E_1 \underset{(M,x)}{\sqcup} E_2 = \langle E_1, E_2 : \sum_j e_{ij}e'_{ij} = 0 \text{ if } \sum_j e_{ij}\, x e_{ij} = 0\rangle$$

The first question to arise is whether this ring is not the trivial ring. We shall show that inside this ring $(E_1E_2,1)$ is isomorphic to (M,x) and so, it can only be trivial if M is. Further, we shall show that $E_1 \underset{(M,x)}{\sqcup} E_2$ is a fir. Therefore, it has a universal skew field of fractions, which we shall denote by $E_1 \underset{(M,x)}{\circ} E_2$.

It is fairly clear that the isomorphism class of the ring $E_1 \underset{(M,x)}{\sqcup} E_2$ depends in general on the generator x that we choose; surprisingly, this is not true for the universal skew field of fractions; $E_1 \underset{(M,x)}{\circ} E_2$ is actually independent up to isomorphism of the generator x, so that it makes sense to talk of the skew field $E_1 \underset{M}{\circ} E_2$ where M is a cyclic bimodule, though in this notation there are no specific embeddings of E_1 and E_2 in $E_1 \underset{M}{\circ} E_2$, whilst there are specified embeddings of E_1 and E_2 into $E_1 \underset{(M,x)}{\circ} E_2$. This result allows us to prove a number of interesting isomorphism theorems.

The method that we develop applies just as well to simple artinian rings S_1 and S_2 in place of E_1 and E_2, so our policy will be to work in this generality whilst pointing out what occurs in the skew field case.

Let S_1 and S_2 be a couple of simple artinian rings, and let (M,x) be a pointed cyclic S_1, S_2 bimodule. A good way to study such a situation is to consider the upper triangular matrix ring $R = \begin{pmatrix} S_1 & M \\ 0 & S_2 \end{pmatrix}$

This ring is a particularly pleasant sort of hereditary ring, so, if we can pull the ring $S_1 \underset{(M,x)}{\sqcup} S_2$ out of it in some way, we shall be able to show that it too has good properties.

Theorem 13.1 Let M be a pointed S_1, S_2 cyclic bimodule, where S_i are both simple artinian rings; let $R = \begin{pmatrix} S_1 & M \\ 0 & S_2 \end{pmatrix}$ and let $\alpha: \begin{pmatrix} S_1 & 0 \\ 0 & 0 \end{pmatrix} \to \begin{pmatrix} 0 & M \\ 0 & S_2 \end{pmatrix}$ be right multiplication by $\begin{pmatrix} 0 & x \\ 0 & 0 \end{pmatrix}$. Then the universal localisation of R at α is isomorphic to $M_2(S_1 \underset{(M,x)}{\sqcup} S_2)$. If $R' = \begin{pmatrix} S_1 & M \oplus N \\ 0 & S_2 \end{pmatrix}$ where N is some S_1, S_2 bimodule, then the universal localisation of R' at α is $M_2(T)$, where T is the tensor ring over $S_1 \underset{(M,x)}{\sqcup} S_2$ on the bimodule $(S_1 \underset{(M,x)}{\sqcup} S_2) \otimes_{S_1} N \otimes_{S_2} (S_1 \underset{(M,x)}{\sqcup} S_2)$. Both of these ring constructions are hereditary.

Proof: In the ring R_α, the elements $\begin{pmatrix} 1 & 0 \\ 0 & 0 \end{pmatrix}$ and $\begin{pmatrix} 0 & 0 \\ 0 & 1 \end{pmatrix}$ and those representing α and its inverse form a set of 2 by 2 matrix units; therefore $R_\alpha \cong M_2(\bar{R})$, where $\bar{R}$ is $\begin{matrix} 1 & 0 \\ 0 & 0 \end{matrix} R_\alpha \begin{matrix} 1 & 0 \\ 0 & 0 \end{matrix}$. $\bar{R}$ is generated by S_1 and $\alpha S_2 \alpha^{-1}$, and the only relations arise because $(S_1 \alpha S_2 \alpha^{-1}, 1)$ as pointed S_1, $\alpha S_2 \alpha^{-1}$ bimodule arises as a quotient of $(S_1 x S_2, x)$ (once we identify S_2 and $\alpha S_2 \alpha^{-1}$ suitably); therefore $\bar{R}$ is isomorphic to $S_1 \underset{(M,x)}{\sqcup} S_2$. Of course, we do not know that $\bar{R}$ is not the trivial ring.

If $R' = \begin{pmatrix} S_1 & M \oplus N \\ 0 & S_2 \end{pmatrix}$, then $R'_\alpha \cong M_2(\bar{R}')$ where $\bar{R}' = \begin{pmatrix} 1 & 0 \\ 0 & 0 \end{pmatrix} R'_\alpha \begin{pmatrix} 1 & 0 \\ 0 & 0 \end{pmatrix}$; $\bar{R}'$ is generated by S_1, $\alpha S_2 \alpha^{-1}$, and $N\alpha^{-1}$, where the only relations that occur state that S_1 and $\alpha S_2 \alpha^{-1}$ generate a copy of $S_1 \underset{(M,x)}{\sqcup} S_2$ and that $N\alpha^{-1}$ is isomorphic as left S_1, right $\alpha S_2 \alpha^{-1}$ bimodule to N as S_1, S_2 bimodule (again we identify S_2 and $S_2 \alpha^{-1}$). This is clearly just the tensor ring over $S_1 \underset{(M,x)}{\sqcup} S_2$ on the bimodule $(S_1 \underset{(M,x)}{\sqcup} S_2) \otimes_{S_1} N \otimes_{S_2} (S_1 \underset{(M,x)}{\sqcup} S_2)$.

Since $M_2(S_1 \underset{(M,x)}{\sqcup} S_2)$ arises as a universal localisation of the hereditary ring R, it must itself be hereditary by 4.9. So $S_1 \underset{(M,x)}{\sqcup} S_2$ is hereditary, and similarly, our tensor ring is hereditary.

When considering an upper triangular matrix ring of the form

$\begin{pmatrix} S_1 & M \\ 0 & S \end{pmatrix}$ where S_1 and S_2 are simple artinian rings, we shall call the projective rank function defined by setting $\rho\begin{pmatrix} S_1 & 0 \\ 0 & 0 \end{pmatrix} = \frac{1}{2} = \rho\begin{pmatrix} 0 & M \\ 0 & S_2 \end{pmatrix}$ the <u>standard u.t. rank function</u>; we have singled this rank function out solely because we shall use it most often, not because it has any special significance. If the map α given by right multiplication by $\begin{pmatrix} 0 & \alpha \\ 0 & 0 \end{pmatrix}$ from $\begin{pmatrix} S_1 & 0 \\ 0 & 0 \end{pmatrix}$ to $\begin{pmatrix} 0 & M \\ 0 & S_2 \end{pmatrix}$ is a full map with respect to this rank function, then R_α will inherit many of the good properties of R as we shall now show. In most applications, it will be clear that this map is full; for example, this is true if S_1 or S_2 is a skew field.

<u>Theorem 13.2</u> Let (M,x) be a pointed cyclic S_1, S_2 bimodule, and let N be some other S_1, S_2 bimodule. Let $R = \begin{pmatrix} S_1 & M\oplus N \\ 0 & S_2 \end{pmatrix}$, and let ρ be the standard u.t. rank function. Then, if right multiplication by $\begin{pmatrix} 0 & \alpha \\ 0 & 0 \end{pmatrix}$ defines a full map from $\begin{pmatrix} S_1 & 0 \\ 0 & 0 \end{pmatrix}$ to $\begin{pmatrix} 0 & M\oplus N \\ 0 & S_2 \end{pmatrix}$, the tensor ring, T, on the bimodule $(S_1 \underset{(M,x)}{\sqcup} S_2) \otimes_{S_1} N \otimes_{S_2} (S_1 \underset{(M,x)}{\sqcup} S_2)$ over the ring $S_1 \underset{(M,x)}{\sqcup} S_2$ is an hereditary ring with a unique projective rank function whose image is generated by the images of the projective rank functions on the subrings S_1 and S_2. In particular, this applies when N is 0; we find that the ring $S_1 \underset{(M,x)}{\sqcup} S_2$ has these properties.

Proof: We know that $M_2(T)$ is the universal localisation of R at the map given by right multiplication by $\begin{pmatrix} 0 & \alpha \\ 0 & 0 \end{pmatrix}$ from $\begin{pmatrix} S_1 & 0 \\ 0 & 0 \end{pmatrix}$ to $\begin{pmatrix} 0 & M\oplus N \\ 0 & S_2 \end{pmatrix}$, which we are assuming is full. $K_0(R) \cong \mathbb{Z} \oplus \mathbb{Z}$, and there is a non-trivial kernel in the map from $K_0(R)$ to $K_0(M_2(T))$, so that there is at most one partial rank function defined on the image of $K_0(R)$ in $K_0(M_2(T))$ so that any projective rank function on $M_2(T)$ must agree with ρ on $K_0(R)$; by the remarks after 5.2, the rank function must be unique, if it exists and there is an extension since α is a full map.

In the case described by the conditions of 13.2, we deduce that the universal localisation of the ring $S_1 \underset{(M,x)}{\sqcup} S_2$ must be simple artinian by 5.5. We denote this by $S_1 \underset{(M,x)}{\circ} S_2$, the universal simple artinian amalgam of S_1 and S_2 along the pointed cyclic bimodule (M,x). Also, we see that $(S_1 S_2, 1)$ is isomorphic to (M,x) as pointed bimodule; for $\begin{pmatrix} S_1 & M \\ 0 & S_2 \end{pmatrix}$ embeds in the universal localisation at the rank function, so it certainly embeds in $\begin{pmatrix} S_1 & M \\ O & S_2 \end{pmatrix}_\alpha$, but we have an isomorphism $(S_1\alpha S_2\alpha^{-1},1) \cong (S_1xS_2,x) \cong (M,x)$, which is what we wanted. We can improve our result in the case where S_1 and S_2 are skew fields.

<u>Theorem 13.3</u> Let E_1 and E_2 be skew fields, and let (M,x) be a pointed cyclic E_1, E_2 bimodule; then $E_1 \underset{(M,x)}{\sqcup} E_2$ is a fir. The pointed cyclic bimodule $(E_1E_2,1)$ is isomorphic to (M,x).

Proof: The map α given by right multiplication by $\begin{pmatrix} O & x \\ O & O \end{pmatrix}$ from $\begin{pmatrix} E_1 & O \\ O & O \end{pmatrix}$ to $\begin{pmatrix} O & M \\ O & E_2 \end{pmatrix}$ is full with respect to the standard u.t. rank function on $R = \begin{pmatrix} E_1 & M \\ O & E_2 \end{pmatrix}$ and it is factor complete by 5.14; so all f.g. projectives over $M_2(E_1 \underset{(M,x)}{\sqcup} E_2) = R$ are induced from R. Since $\begin{pmatrix} E_1 & O \\ O & O \end{pmatrix}$ and $\begin{pmatrix} O & M \\ O & E_2 \end{pmatrix}$ become isomorphic over the universal localisation at α, it follows that $E_1 \underset{(M,x)}{\sqcup} E_2$ must be a fir.

<u>Isomorphism theorems</u>

<u>Theorem 13.4</u> Let S_1 and S_2 be simple artinian rings, and let M be a cyclic S_1, S_2 bimodule. Assume that x and y are both generators of M as a bimodule such that $\begin{pmatrix} O & x \\ O & O \end{pmatrix}$ and $\begin{pmatrix} O & y \\ O & O \end{pmatrix}$ both define full maps from $\begin{pmatrix} S_1 & O \\ O & O \end{pmatrix}$ to $\begin{pmatrix} O & M \\ O & S_2 \end{pmatrix}$ on the ring $\begin{pmatrix} S_1 & M \\ O & S_2 \end{pmatrix}$ with respect to the standard u.t. rank function. Then $S_1 \underset{(M,x)}{\circ} S_2$ is isomorphic to $S_1 \underset{(M,x)}{\circ} S_2$.

Proof: The universal localisation of $\begin{matrix} S_1 & M \\ O & S \end{matrix}$ at the given rank function is both the universal localisation of $M_2(S_1 \underset{(M,x)}{\sqcup} S_2)$ at the unique rank function and the universal localisation of $M_2(S_1 \underset{(M,y)}{\sqcup} S_2)$ at its rank function; so, $M_2(S_1 \underset{(M,x)}{\circ} S_2) \cong M_2(S_1 \underset{(M,y)}{\circ} S_2)$; since both sides of the isomorphism are simple artinian rings, we can deduce that $S_1 \underset{(M,x)}{\circ} S_2 \cong S_1 \underset{(M,y)}{\circ} S_2$.

For skew fields, we can eliminate a hypothesis from this theorem.

<u>Theorem 13.5</u> Let E_1 and E_2 be skew fields; let M be a cyclic E_1, E_2 bimodule; then $E_1 \underset{(M,x)}{\circ} E_2 \cong E_1 \underset{(M,y)}{\circ} E_2$ for any bimodule generators x

and y of M.

If $y = \Sigma e_i x e_i'$, it is possible to calculate that a specific map from $E_1 \underset{(M,y)}{\sqcup} E_2$ to $E_1 \underset{(M,x)}{\circ} E_2$ that extends to an isomorphism of $E_1 \underset{(M,y)}{\circ} E_2$ with $E_1 \underset{(M,x)}{\circ} E_2$ is given by $E_2 \to E_2$ by the identity map, and E_1 is mapped to $E_1^{\Sigma e_i e_i'}$.

We see from the last theorem that if F_1 and F_2 are common skew subfields of E_1 and E_2 such that $E_1 \otimes_{F_1} E_2 \cong E_1 \otimes_{F_2} E_2$ as E_1, E_2 bimodules, then $E_1 \underset{F_1}{\circ} E_2 \cong E_1 \underset{F_2}{\circ} E_2$. Our next result simply states some interesting special cases of this result.

<u>Theorem 13.6</u> Let S_1 be a f.d. simple artinian k-algebra, and let S_2 be a simple artinian k-algebra such that $S_1^{o} \otimes_k S_2$ is simple artinian. If T_1 and T_2 are common f.d. simple artinian k-algebras such that $[T_1:k] = [T_2:k]$, then $S_1 \underset{T_1}{\circ} S_2 \cong S_1 \underset{T_2}{\circ} S_2$.

Proof: $S_1 \otimes_{T_1} S_2 \cong S_1 \otimes_{T_2} S_2$, since the ring $S_1^{o} \otimes_k S_2$ is simple artinian, and both bimodules are free S_2 modules of the same rank. So

$$R = \begin{pmatrix} S_1 & S_1 \otimes_{T_1} S_2 \\ 0 & S_2 \end{pmatrix} \cong \begin{pmatrix} S_1 & S_1 \otimes T_2 S_2 \\ 0 & S_2 \end{pmatrix}.$$

Right multiplication by the element $\begin{pmatrix} 0 & 1 \otimes_{T_1} 1 \\ 0 & 0 \end{pmatrix}$ defines a full map with respect to the standard u.t. rank function from $R\begin{pmatrix} 1 & 0 \\ 0 & 0 \end{pmatrix}$ to $R\begin{pmatrix} 0 & 0 \\ 0 & 1 \end{pmatrix}$, since the localisation of R is isomorphic to $M_2(S_1 \underset{T_1}{\sqcup} S_2)$ by 13.2, which has a homomorphism to a simple artinian ring inducing our rank function on R. Similarly, the element $\begin{pmatrix} 0 & 1 \otimes_{T_2} 1 \\ 0 & 0 \end{pmatrix}$ defines a full map; therefore, by 13.4, $S_1 \underset{T_1}{\circ} S_2 \cong S_1 \underset{T_2}{\circ} S_2$.

Our last result gives us a number of isomorphic simple artinian coproducts with the same factors but different amalgamated simple artinian subrings; our next result uses this to provide examples of isomorphic simple artinian coproducts that have different factors and the amalgamation takes place over the same central subfield.

<u>Theorem 13.7</u> Let S be a f.d. simple artinian k-algebra, and let S_1 and S_2 be simple artinian subalgebras of S such that $[S_1:k] = [S_2:k]$; then $S \underset{k}{\circ} S_1 \cong S \underset{k}{\circ} S_2$.

Proof: $S \underset{k}{\circ} S_1 \cong S \underset{S_2}{\circ} (S_2 \underset{k}{\circ} S_1) \cong S \underset{S_1}{\circ} (S_2 \underset{k}{\circ} S_1)$ by 13.6, since the centre of $S_1 \underset{k}{\sqcup} S_2$ is k or else purely transcendental over of degree 1, as we saw in chapter 10, so that $S^o \otimes k(S_1 \underset{k}{\circ} S_2)$ is simple artinian. However, $S \underset{S_1}{\circ} (S_2 \underset{k}{\circ} S_1) \cong S \underset{k}{\circ} S_2$.

We can prove a stronger theorem than 13.7 in the case where S is a central simple artinian k-algebra, that is, when the centre of S is exactly k.

<u>Theorem 13.8</u> Let S be a central simple k-algebra such that $[S:k] = n^2$; let S_1 be a simple artinian k-subalgebra such that $[S_1:k] = m$. Then $S \underset{k}{\circ} S_1 \cong S \otimes_k k\langle X\rangle$, where $|X| = n^2 \frac{m-1}{m}$.

Proof: $M_2(S \underset{k}{\circ} S_1)$ is the universal localisation of $\begin{pmatrix} S & S\otimes_k S_1 \\ 0 & S_1 \end{pmatrix}$ at the standard u.t. rank function. As a bimodule, $S\otimes_k S_1 \cong \bigoplus_{i=1}^{m} S$, m copies of the bimodule S where S acts on the left in the obvious way, and S_1 acts on the right via the embedding of S_1 in S.

By 13.1, the universal localisation of $\begin{pmatrix} S & \bigoplus_{i=1}^{m} S \\ 0 & S_1 \end{pmatrix}$ at the map from $\begin{pmatrix} S & 0 \\ 0 & 0 \end{pmatrix}$ to $\begin{pmatrix} 0 & \oplus S \\ 0 & S_1 \end{pmatrix}$ given by right multiplication by $\begin{pmatrix} 0 & (1,0,0\ldots) \\ 0 & 0 \end{pmatrix}$ is $M_2(T)$, where T is the tensor ring over $S \underset{S_1}{\circ} S_1 \cong S$ on the bimodule $\bigoplus_{i=1}^{m} S\otimes_S S\otimes_{S_1} S$, which is $\bigoplus_{i=1}^{m} S\otimes_{S_1} S$. This is the direct sum of $n^2 \frac{m-1}{m}$ simple bimodules; so $T \cong S\langle X\rangle$ where X is a set of $n^2 \frac{m-1}{m}$ elements which we identify with $n^2 \frac{m-1}{m}$ S-centralising generators of this bimodule. Hence, the universal localisation of T at its unique rank function is $S\otimes_k k\langle X\rangle$.

Therefore, $M_2(S\otimes_k k\langle X\rangle)$ is a universal localisation of $\begin{pmatrix} S & S\otimes_k S_1 \\ 0 & S_1 \end{pmatrix}$ that is simple artinian, and the rank function induced on this ring is ρ; so $S\otimes_k k\langle X\rangle \cong S \underset{k}{\circ} S_1$.

We can use our methods in a rather more complicated way in order to investigate simple artinian coproducts of the form $M_m(k) \underset{k}{\circ} M_n(k)$; we should like to show that these are all suitably sized matrix rings over a free skew field on a suitable number of generators. We are not able to deal with this degree of generality at present, but there are a number of special cases of interest where we can prove this result; we shall also show that they are all suitably sized matrix rings over stably free skew fields. A skew field, F, is said to be <u>stably free</u>, if $F \underset{k}{\circ} k\langle X\rangle \cong k\langle Y\rangle$, for a finite set X. Of course, if m divides n, our last theorem applies to show that $M_m(k) \underset{k}{\circ} M_n(k) \cong M_n(k\langle X\rangle)$ for suitable X; this actually allows us to prove the next theorem with little effort. Before we do this, we note a few generalities. We write $M_m(k) \underset{k}{\circ} M_n(k) \cong M_p(F)$; since the rank function on $M_m(k) \underset{k}{\sqcup} M_n(k)$ has as image $\frac{1}{p}\mathbb{Z}$ where p is the least common multiple of m and n, and $M_p(F)$ is the universal localisation of $M_m(k) \underset{k}{\sqcup} M_n(k)$ at this rank function, it follows that p is the least common multiple of m and n.

<u>Theorem 13.9</u> Suppose that $M_s(k) \underset{k}{\circ} M_t(k) \cong M_p(F)$, where $p = \text{l.c.m.}\{s,t\}$; then $M_{sn}(k) \underset{k}{\circ} M_{tn}(k) \cong M_{pn}(F \underset{k}{\circ} k\langle Y\rangle)$ for a suitable finite set Y.

Proof: $M_{sn}(k) \underset{k}{\circ} M_n(k) \cong M_{sn}(k\langle Z_1\rangle)$ for some set Z_1 by 13.8. So

$$M_{sn}(k) \underset{k}{\circ} M_{tn}(k) \cong M_{sn}(k) \underset{k}{\circ} M_n(k) \underset{M_n(k)}{\circ} M_{tn}(k) \cong M_{sn}(k\langle Z_1\rangle) \underset{M_n(k)}{\circ} M_{tn}(k);$$

this is isomorphic to $M_n(M_s(k\langle Z_1\rangle) \underset{k}{\circ} M_t(k))$ as we see by taking the centraliser of $M_n(k)$ inside it.

$$M_s(k\langle Z_1\rangle) \underset{k}{\circ} M_t(k) \cong M_s(k\langle Z\rangle) \underset{M_s(k)}{\circ} M_s(k) \underset{k}{\circ} M_t(k)$$

$\cong M_s(k\langle Z_1\rangle) \underset{M_s(k)}{\circ} M_p(F)$. Taking the centraliser of $M_s(k)$, we find that this is the ring $M_s(k\langle Z_1\rangle \circ_k M_{p'}(F))$, where $p' = p/s$.

$k\langle Z_1\rangle \circ_k M_{p'}(F)$ is the universal localisation of the hereditary ring $k<Z_1> \sqcup_k M_{p'}(F) \cong k<Z_1> \sqcup_k M_{p'}(k) \sqcup_{M_{p'}(k)} M_{p'}(F) \cong M_{p'}(k<Z_2>) \sqcup_{M_{p'}(k)} M_{p'}(F)$, where Z_2 is the set of components of elements of Z_1 written as matrices over the centraliser of $M_{p'}(k)$ in $M_{p'}(k) \sqcup_k k<Z_1>$. It is clear that this ring is just $M_{p'}(F \sqcup_k k<Z_2>)$, which is a hereditary ring whose only universal localisation that is simple artinian is $M_{p'}(F \circ_k K\langle Z_2\rangle)$. Tracing the argument back shows that $M_{sn}(k) \circ_k M_{tn}(k) \cong M_{pn}(F \circ_k k\langle Z_2\rangle)$.

In order to go a little further, we look at suitable universal localisations of the ring $R = \begin{pmatrix} M_m(k) & M_m(k)\otimes_k M_n(k) \\ 0 & M_n(k) \end{pmatrix}$, which is a f.d. hereditary k-algebra; we shall use 13.7 to find good universal localisations of related rings, which allow us under suitable restrictions on m and n to construct universal localisations of R at full maps with respect to the rank function assigning the rank $\frac{1}{2}$ to both indecomposable projectives, that have the form $M_{2p}(k\langle X\rangle)$.

We recall from the last chapter that the epimorphism from a f.d. hereditary algebra, R, associated to a pre-projective or pre-injective module right module is always a universal localisation. For our purposes, we take R to be the ring $\begin{pmatrix} k & k^s \\ 0 & k \end{pmatrix}$, where $s \geq 2$. We wish to find the rank function associated to a particular module M. This is given by $\rho(\begin{pmatrix} 1 & 0 \\ 0 & 0 \end{pmatrix}R) = [M\begin{pmatrix} 1 & 0 \\ 0 & 0 \end{pmatrix}:k]/[M:k]$, and $\rho(\begin{pmatrix} 0 & 0 \\ 0 & 1 \end{pmatrix}R) = [M\begin{pmatrix} 0 & 0 \\ 0 & 1 \end{pmatrix}:k]/[M:k]$. We summarise from Dlab, Ringel (76), what rank functions occur associated to pre-projective or pre-injective modules over R; the pairs $([M\begin{pmatrix} 1 & 0 \\ 0 & 0 \end{pmatrix}:k],[M\begin{pmatrix} 0 & 0 \\ 0 & 1 \end{pmatrix}:k])$ that occur are the positive real roots associated to the graph, , where there are s arrows from the first point to second. Rather than developing here the precise theory which gives us these positive real roots, we give an ad hoc description of them. We begin with the pair $(0,1)$; our inductive procedure is to pass from the pair (a,b)

to the pair $(b, sb - a)$; finally, if (a,b) is a pair constructed above, then (b,a) is also a pair.

If (a,b) is a particular positive root corresponding to a module M, then $[M:k] = a + b$, so that the rank function associated to the epimorphism given by M is $\rho\left(\begin{pmatrix}1 & 0\\ 0 & 0\end{pmatrix}R\right) = \frac{a}{a+b}$, $\rho\left(\begin{pmatrix}0 & 0\\ 0 & 1\end{pmatrix}R\right) = \frac{b}{a+b}$. That is, we have an epimorphism from $\begin{matrix}k & k^s\\ 0 & k\end{matrix}$ to $M_{a+b}(k)$ which must be a universal localisation at the above rank function.

By Morita equivalence, we must have an epimorphism from $\begin{pmatrix}M_b(k) & S^s\\ 0 & M_a(k)\end{pmatrix}$ to $M_{2ab}(k)$, which is a universal localisation. Here, S is the unique simple left $M_b(k)$, right $M_a(k)$ bimodule. The rank function associated to this universal localisation is the standard u.t. rank function. Consequently, we consider the ring $\begin{pmatrix}M_b(k) & M_b(k)\otimes_k M_a(k)\\ 0 & M_a(k)\end{pmatrix}$ as

$$\begin{pmatrix}M_b(k) & S^s \oplus S^{ab-s}\\ 0 & M_a(k)\end{pmatrix}$$

We see from the above that we have a universal localisation of this which by 13.1, has the form $M_2(T)$, where $T \cong M_{ab}(k)\langle (M_{ab}(k) \otimes_{M_b(k)} S^{ab-s} \otimes_{M_a(k)} M_{ab}(k)\rangle$. A dimension check shows that the bimodule in the brackets is isomorphic to $M_{ab}(k)^{ab(ab-s)}$ as $M_{ab}(k)$ bimodule. Consequently, $T \cong M_{ab}(k\langle X\rangle)$, where $|X| = ab(ab-s)$.

Therefore, the universal localisation of $\begin{pmatrix}M_b(k) & M_b(k)\otimes_k M_a(k)\\ 0 & M_a(k)\end{pmatrix}$ at the rank function we are considering is $M_{2ab}(k\langle X\rangle)$. However, so is $M_2(M_b(k) \underset{k}{\circ} M_a(k))$; therefore, $M_a(k) \underset{k}{\circ} M_b(k) \cong M_{ab}(k\langle X\rangle)$.

We summarise what we have shown in the next theorem.

<u>Theorem 13.10</u> Let (a,b) be a positive real root associated to the graph

$\cdot \overset{\rightarrow}{\underset{\rightarrow}{\vdots}} \cdot$, where there are s arrows from the first point to the second. Then

$M_a(k) \underset{k}{\circ} M_b(k) \cong M_{ab}(k\langle X\rangle)$, where $|X| = ab(ab-s)$.

If $s = 2$, the pairs (a,b) that occur as positive roots are exactly the pairs $(n,n+1)$ or $(n+1,n)$, so that $M_n(k) \circ_k M_{n+1}(k) \cong M_{n(n+1)}(k\langle X\rangle)$ for a suitable finite set X. We use this to prove the last result of this section.

<u>Theorem 13.11</u> Let m and n be arbitrary positive integers; then $M_m(k) \circ_k M_n(k) \cong M_p(F)$, where p is the least common multiple of m and n, and F is a stably free skew field.

Proof: By 13.9, we may assume that m and n are co-prime, since we may reduce to this case. Let s and t be integers such that $sm = tn + 1$. We know that $M_{mst}(k) \circ_k M_{nst}(k) \cong M_{mnst}(F \circ_k k\langle Y\rangle)$ for some finite set Y by 13.9.

But $M_{mst}(k) \circ_k M_{nst}(k) \cong M_{mst}(k) \circ_{M_{ms}(k)} M_{ms}(k) \circ_k M_{nt}(k) \circ_{M_{nt}(k)} M_{nst}(k)$

which is isomorphic to $M_{mst}(k) \circ_{M_{ms}(k)} M_{mnst}(k\langle Z\rangle) \circ_{M_{nt}(k)} M_{nst}(k)$ by 13.10,

for a finite set Z, since $sm = tn + 1$. If a divides b, $M_a(k) \circ_k M_b(D) \cong M_a(k) \circ_k M_b(k) \circ_{M_b(k)} M_b(D)$ which is $M_b(k\langle X\rangle) \circ_{M_b(k)} M_b(D)$, by 13.8; this is isomorphic to $M_b(D \circ_k k\langle X\rangle)$. So, applying this twice completes the proof that $M_{mst}(k) \circ_k M_{nst}(k) \cong M_{mnst}(k\langle Z'\rangle)$, where Z' is a finite set.

We have no examples of a skew subfield of a free skew field that is not itself free, so certainly, we have no examples of stably free skew fields that are not free. At present, it is not clear how these matters will eventually settle themselves.

<u>Artin's problem for skew field extensions</u>

The purpose of this section is to construct extensions of skew fields $E \supset F$ such that the left dimension of E over F, $[E:F]_l$, and the right dimension of E over F, $[E:F]_r$, are an arbitrary pair of integers greater than 1. In fact, we shall consider a slightly more general problem; what are the possible pairs of integers for the left and right dimension of a simple artinian ring $M_n(E)$ over a skew subfield F? We shall see that arbitrary pairs of integers greater than 1 and divisible by n occur; that the left and right dimensions are divisible by n follows from the observation that $M_n(E)$ as a left or right module over itself is the direct summand of n isomorphic simple modules.

We shall actually prove something rather stronger than this, and the extra strength will be important in the final section where we shall construct hereditary artinian rings of representation type $I_2(5)$. In order to set up the notation used throughout this section, we state precisely what we shall prove. Throughout this section, the skew fields E and F, the elements $\{e_{ij}: i = 1 \text{ to } n\}$, the elements $\{s_{kj}: k = 1 \text{ to } a,\ j = 1 \text{ to } n\}$, the elements $\{t_{ih}: i = 1 \text{ to } n \ \ h = 1 \text{ to } b\}$ and the integers n, a and b will be named as in the following theorem.

Theorem 13.12 Let $M_n(E) \supset F$, where E and F are skew fields. Let $\{e_{ij}: i,j = 1 \text{ to } n\}$ be a set of matrix units in $M_n(E)$. Let a,b be integers such that $an, bn > 1$: let $\{s_{kj}: k = 1 \text{ to } a,\ j = 1 \text{ to } n\}$ be elements of $M_n(E)$ that are left independent over F such that $s_{kj}e_{jj} = s_{kj}$ and let $\{t_{ih}: i = 1 \text{ to } n,\ h = 1 \text{ to } b\}$ be elements of $M_n(E)$ that right independent over F such that $e_{ii}t_{ih} = t_{ih}$. Then there exist skew fields $\bar{E} \supset E$, $\bar{F} \supset F$ and a diagram of rings:

$$\begin{array}{ccc} M_n(\bar{E}) & \supset & M_n(E) \\ \cup & & \cup \\ \bar{F} & \supset & F \end{array}$$

where $\bar{F} \cap M_n(E) = F$, a left basis for $M_n(E)$ over F is $\{s_{kj}\}$ and a right basis for $M_n(E)$ over F is $\{t_{ih}\}$.

In order to construct E and F in theorem 13.12, we shall need an intermediate construction.

Theorem 13.13 Given the data of theorem 13.12, there exist skew fields E' and F' and a diagram of rings:

$$\begin{array}{ccc} M_n(E') & \supset & M_n(E) \\ \cup & & \cup \\ F' & \supset & F \end{array}$$

where $F' \cap M_n(E) = F$, a basis for $F'M_n(E)$ over F' is $\{s_{kj}\}$ whilst the pointed bimodule $(M_n(E)F', 1)$ is isomorphic to $(M_n(E) \otimes_F F', 1 \otimes 1)$.

Notice that the extension of rings $M_n(E') \supset F'$ still satisfies all the conditions originally stated for the extension of rings $M_n(E) \supset F$.

There is also a version of theorem 13.13 interchanging the role of left and right, of a and b, and of $\{s_{kj}\}$ and $\{t_i\}$, which we leave to the reader to formulate; we shall refer to this as theorem 13.13' when we need to mention it.

We begin by showing how theorem 13.12 follows from 13.13 and 13.13'.

We begin with the extension of rings $M_n(E) \supset F$; for inductive purposes we set $E = E_0$ and $F = F_0$.

At an odd stage in our construction we assume that we have a diagram of rings:

$$\begin{array}{ccc} M_n(E_{2m}) & \supset & F_{2m} \\ \cup & & \cup \\ M_n(E_0) & \supset & F_0 \end{array}$$

such that $F_{2m} \cap M_n(E_0) = F_0$ and E_{2m}, F_{2m} are skew fields such that the conditions of theorem 13.12 are satisfied on replacing E and F by E_{2m} and F_{2m}; by theorem 13.13, we construct skew fields E_{2m+1}, F_{2m+1} with a diagram of rings:

$$\begin{array}{ccc} M_n(E_{2m+1}) & \supset & F_{2m+1} \\ \cup & & \cup \\ M_n(E_{2m}) & \supset & F_{2m} \end{array}$$

such that $F_{2m+1} \cap M_n(E_{2m}) = F_{2m}$, a left basis for $F_{2m+1}M_n(E_{2m})$ over F_{2m+1} is $\{s_{kj}\}$ whilst $(M_n(E_{2m})F_{2m+1}, 1) \cong (M_n(E_{2m}) \otimes_{F_{2m}} F_{2m+1}, 1\otimes 1)$ as pointed bimodule. It follows also that $M_n(E_0) \cap F_{2m+1} = F_0$ and that the conditions of theorem 13.12 are satisfied with E_{2m+1} and F_{2m+1} replacing E and F.

At an even stage of the construction, we assume that we have an extension of rings:

$$\begin{array}{ccc} M_n(E_{2m-1}) & \supset & F_{2m-1} \\ \cup & & \cup \\ M_n(E_0) & \supset & F_0 \end{array}$$

such that $F_{2m-1} \cap M_n(E_0) = F_0$ where E_{2m-1} and F_{2m-1} are skew fields such that the conditions of theorem 13.12 are satisfied on replacing E and F by E_{2m-1} and F_{2m-1}; using theorem 13.13', we construct skew fields E_{2m} and F_{2m} with a diagram of rings:

$$\begin{array}{ccc} M_n(E_{2m}) & \supset & F_{2m} \\ \cup & & \cup \\ M_n(E_{2m-1}) & \supset & F_{2m-1} \end{array}$$

such that $M_n(E_{2m-1}) \cap F_{2m} = F_{2m-1}$, a right basis for $M_n(E_{2m-1})F_{2m}$ over F_{2m} is $\{t_{ih}\}$ whilst $(F_{2m}M_n(E_{2m-1}),1) \cong (F_{2m} \otimes_{F_{2m-1}} M_n(E_{2m-1}), 1\otimes 1)$ as pointed bimodule. It follows that $M_n(E_0) \cap F_{2m} = F_0$ and that the conditions of theorem 13.12 are satisfied with E_{2m} and F_{2m} replacing E and F.

On setting $\bar{E} = \cup_i E_i$, $F = \cup_i F_i$, we have an extension of rings $M_n(\bar{E}) \supset \bar{F}$ such that $M_n(E) \cap \bar{F} = F$. If $s \in M_n(E)$, it lies in $M_n(E_{2p})$ for some integer p; consequently it is left dependent on $\{s_{kj}\}$ over F_{2p+1} and right dependent on $\{t_{ih}\}$ over F_{2p+2}. If there is a dependence relation between the elements $\{s_{kj}\}$ on the left over $\bar{F}$, the dependence relation actually occurs over F_q for some q; however, this cannot happen by construction; similarly, the elements $\{t_{ih}\}$ remain right independent over F. Therefore, the elements $\{s_{kj}\}$ are a left basis for $M_n(E)$ over F, and the elements $\{t_{ih}\}$ form a right basis, which completes the proof of theorem 13.12 assuming that we can prove theorem 13.13.

It remains to be seen how to prove theorem 13.13. There are two problems to overcome; firstly, we must construct a skew field $F' \supset F$ and an F', $M_n(E)$ bimodule M with an F-centralising generator x such that $[M:F'] = an$ with basis $\{xs_{kj}\}$; secondly, we must construct a simple artinian ring $M_n(E')$ containing F' and $M_n(E)$ such that $(F'M_n(E),1) \cong (M,x)$ as pointed F', $M_n(E)$ bimodule. We have already seen an approach to the second construction; we set $M_n(E') = F' \circ_{(M,x)} M_n(E)$; for the first construction, we set $M_{an}(F') = M_{an}(F) \circ_F M_n(E)$; let $g_{\alpha\beta}$ be matrix units where α and β run over pairs kj for $k = 1$ to a and $j = 1$ to n; let $M = g_{11\,11}M_{an}(F')$ and $x = g_{11\,11}$; we have yet to show that x is a generator of the bimodule M. Certainly, $[M:F'] = an$, and we intend to show that $\{xs_{kj}\}$ is a basis for M over F'. For the time being we rename the elements $\{s_{kj}\}$ as the elements s_α since the indexing sets are the same. If the elements $\{xs_\alpha\}$ are not a basis, there is a relation $\Sigma f_\alpha xs_\alpha = 0$ for elements $f_\alpha \in F'$; rewriting this equation in the ring $M_{an}(F')$ we find that the row whose αth entry is f_α kills the matrix which lies in $M_{an}(F) \sqcup_F M_n(E)$, $\sum_\alpha g_{\alpha 11} s_\alpha$. This implies that the element $\sum_\alpha g_{\alpha 11} s_\alpha$ lies in a left ideal of rank less than 1 where the rank is the unique one on $M_{an}(F) \sqcup_F M_n(E)$; in turn, $\begin{pmatrix} 0 & \sum g_{\alpha 11} s_\alpha \\ 0 & 0 \end{pmatrix}$ of $T = \begin{pmatrix} M_{an}(F) & M_{an}(F) \otimes_F M_n(E) \\ 0 & M_n(E) \end{pmatrix}$ must lie in a left ideal of standard u.t. rank $< \frac{1}{2}$. Left ideals of T inside $T\begin{pmatrix} 0 & 0 \\ 0 & 1 \end{pmatrix}$ take the form $\begin{pmatrix} 0 & M_{an}(F) \otimes_F I \\ 0 & I \end{pmatrix} \oplus \begin{pmatrix} 0 & N \\ 0 & 0 \end{pmatrix}$ where I is a left ideal of $M_n(E)$ and N is an $M_{an}(F)$ submodule of $M_{an}(F) \otimes_F M_n(E)$. Its standard u.t. rank is $\frac{1}{2}(\rho(I) + \rho(N))$. Let $J = \begin{pmatrix} 0 & M_{an}(F) \otimes_F I \\ 0 & I \end{pmatrix} \oplus \begin{pmatrix} 0 & N \\ 0 & 0 \end{pmatrix}$ be a left ideal of T containing $\begin{pmatrix} 0 & \sum_\alpha g_{\alpha 11} s_\alpha \\ 0 & 0 \end{pmatrix}$ of standard u.t. rank $< \frac{1}{2}$.

We have a map from $g_{11\,11} \otimes_F \sum_\alpha Fs_\alpha \le M_{an}(F) \otimes_F I \oplus N$ to $g_{11\,11} N$. From the inequality $\rho(I) + \rho(N) < 1$, we deduce that $[g_{11\,11} N:F] < an(1 - \rho(I))$; so we find that $[\sum_\alpha Fs_\alpha \cap I:F] > an\rho(I)$.

It is convenient to return to the kj notation rather than indexing by α.

Since $\rho(I) = m/n < 1$ for some integer m, then for some i, $I \cap M_n(E)e_{ii} = 0$, and so, $I \cap \sum_k Fs_{ki} = 0$; then $\rho(I + M_n(E)e_{ii}) = (m + 1)/n$ and also $[(I + M_n(E)e_{ii}) \cap \sum_{k,j} Fs_{kj}:F] > an\rho(I) + a = an(\rho(I + M_n(E)e_{ii}))$; we set I' to be $I + M_n(E)e_{ii}$ and if $I' \ne M_n(E)$, we repeat the argument; by induction, we eventually find $an = [M_n(E) \cap \sum_{k,j} Fs_{kj}:F] > an\rho(M_n(E)) = an$.

This contradiction shows that the element $\sum g_{\alpha 11} s_\alpha$ is not a zero-divisor in $M_{an}(F')$ and so, the elements $\{xs_{kj}\} = \{g_{11\,11} s_{kj}\}$ form a basis for $M = g_{11\,11} M_{an}(F')$ over F'. So, M is a cyclic F', $M_n(E)$ bimodule on the generator x.

We consider the simple artinian ring $M_n(E') = F' \underset{(M,x)}{\circ} M_n(E)$;

we already know that $(F'M_n(E),1)$ is isomorphic to (M,x) as pointed F', $M_n(E)$ bimodule; so $F'M_n(E)$ has as left basis over F' the set $\{s_{kj}\}$.

Also, $F' \cap M_n(E) = F$; for let $s \in M_n(E)$ such that $xs = fx$ for some element f in F'; rewriting this equation in $M_{an}(F')$ gives $g_{11\,11}s = fg_{11\,11}$; so, inside $M_{an}(F) \mathbin{\square_F} M_n(E)$, $g_{11\,11}sg_{11} = g_{11\,11}s$; it follows that s lies in F and $s = f$. Another way of stating this is that the normaliser of x in M is F.

It remains to show that $(M_n(E)F',1)$ is isomorphic to $(M_n(E)\otimes_F F',1\otimes 1)$; in order to prove this, it is simpler to prove the following more general result.

<u>Theorem 13.14</u> Let R be a semihereditary ring with faithful projective rank function ρ taking values in $\frac{1}{n}\mathbb{Z}$; let $\alpha:P \to Q$ be an atomic full map between f.g. left projectives; let N be the normaliser of α which we regard as a subring of $E(P) = \mathrm{End}_R(P)$ and $E(Q) = \mathrm{End}_R(Q)$ via the embedding as left and right normaliser respectively; then, in the category of f.g. projectives over R_α and of R_ρ, $(E(Q)\alpha^{-1}E(P),\alpha^{-1})$ is isomorphic as pointed bimodule to $(E(Q)\otimes_N E(P),1\otimes 1)$.

Proof: It is immaterial whether we consider the bimodule in the category of f.g. projectives over R_α or over R_ρ since $\{\alpha\}$ is a factor closed set of maps and so by theorem 5.8 R_α embeds in R_ρ.

Certainly, there is a natural map from $E(Q)\otimes_N E(P)$ to $E(Q)\alpha^{-1}E(P)$ sending $1\otimes 1$ to α^{-1}. It remains to check that there is no kernel.

There are two natural maps from $\mathrm{Hom}_R(Q,P)$ to $E(Q)\otimes_N E(P)$; $\phi_1:\beta \to \beta\alpha\otimes 1$ and $\phi_2:\beta \to 1\otimes\alpha\beta$, and their images agree under the natural map from $E(Q)\otimes_N E(P)$ to $E(Q)\alpha^{-1}E(P)$. However, we have the relation $\alpha(\beta\alpha) = (\alpha\beta)\alpha$ from which we see that the element of N that goes to $\beta\alpha$ in $E(Q)$ goes to $\alpha\beta$ in $E(P)$; it follows that the maps ϕ_1 and ϕ_2 agree; it is an embedding since the composition with the map to $E(Q)\alpha^{-1}E(P) \subseteq \mathrm{Hom}_{R_\rho}(R_\rho\otimes_R Q, R_\rho\otimes_R P)$ is the embedding of $\mathrm{Hom}_R(Q,P)$; we shall regard $\mathrm{Hom}_R(Q,P)$ as an $E(Q),E(P)$ sub-bimodule via this map.

Consider a relation in $E(Q)\alpha^{-1}E(P)$;

1/ $$\sum_{i=1}^{n} \beta_i\alpha^{-1}\gamma_i + \delta = 0$$

where $\delta \in \mathrm{Hom}_R(Q,P)$; we shall show by induction on n that the relation

$\sum_{i=1}^{n} \beta_i \otimes \gamma_i + \delta = 0$ holds in $E(Q) \otimes_N E(P)$; so far we have shown this to be true if $n = 0$.

From the relation 1/, we find that the map

$$\left(\begin{array}{cccc|c} \alpha & & & 0 & \gamma_1 \\ & \alpha & & & \gamma_2 \\ & & \ddots & & \vdots \\ 0 & & & \alpha & \gamma_n \\ \hline \beta_1 & \beta_2 & \cdots & \beta_n & -\delta \end{array}\right)$$

has nullity $\rho(P) = \rho(Q)$.

We write an equation expressing this defined over R:

$$\left(\begin{array}{ccc|c} \alpha & & 0 & \gamma_1 \\ & \ddots & & \vdots \\ 0 & & \alpha & \gamma_n \\ \hline \beta_1 & & \beta_n & -\delta \end{array}\right) = \frac{\mu}{\phi_1} (\nu \mid \phi_2)$$

μ and ν must be full maps such that $\mu\nu = \begin{pmatrix} \alpha & & 0 \\ & \ddots & \\ 0 & & \alpha \end{pmatrix}$. By 1.19, there is an invertible map ε such that $\mu\varepsilon = \begin{pmatrix} \mu_1 & & 0 \\ & \ddots & \\ * & & \mu_n \end{pmatrix}$ and $\varepsilon^{-1}\nu = \begin{pmatrix} \nu_1 & & 0 \\ & \ddots & \\ * & & \nu_n \end{pmatrix}$

for full maps μ_i, ν_i we may adjust our previous equation to obtain the equation:

$$\left(\begin{array}{ccc|c} \alpha & & 0 & \gamma_1 \\ & \ddots & & \vdots \\ 0 & & \alpha & \gamma_n \\ \hline \beta_1 & \cdots & \beta_n & \end{array}\right) = \left(\begin{array}{ccc} \mu_1 & & 0 \\ & \ddots & \\ * & & \mu_n \\ \hline \sigma_1 & \cdots & \sigma_n \end{array}\right) \left(\begin{array}{ccc|c} \nu_1 & & 0 & \tau_1 \\ & \ddots & & \vdots \\ * & & \nu_n & \tau_n \end{array}\right)$$

Since α is an atomic full map, we may further assume that one of μ_i, ν_i is an identity map whilst the other is α. If $\mu_1 = \alpha$, $\gamma_1 = \alpha\tau_1$ which lies in N and this allows us to shorten our relation by using a relation holding in $E(Q)\otimes_N E(P)$; by induction we deduce that our original relation is a consequence of relations in $E(Q)\otimes_N E(P)$. Similarly, if μ_n is an identity map and $\nu_n = \alpha$, then $\beta_n = \sigma_n \alpha$ and again our relation is forced to be a consequence of a relation in $E(Q)\otimes_N E(P)$. So, we may assume that μ_1 is the identity map on P and $\mu_n = \alpha$; let μ_{k+1} be the first μ_i that is α. We rewrite our equation with this information:

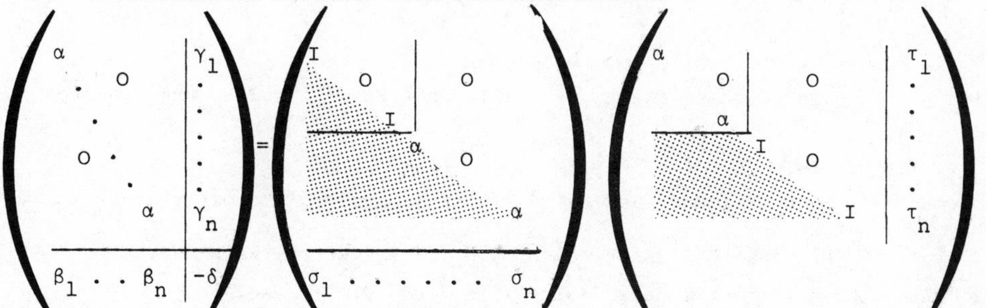

Since the top left hand corner of the leftmost map on the right hand side is invertible, we may adjust our equation again to obtain:

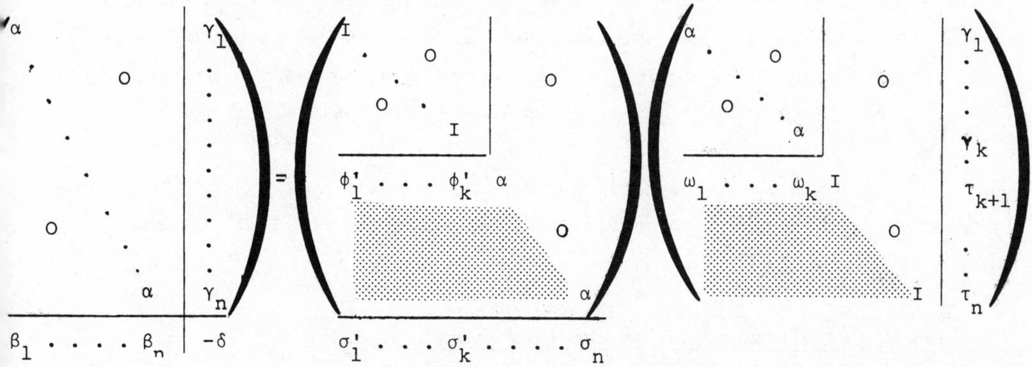

From this, we have the relations $\phi_i\alpha + \alpha\omega_i = 0 : \phi_i \in E(P), \omega_i \in E(Q)$; so ϕ_i and ω_i lie in the left and right normaliser of α respectively. Further, we have the relation $\gamma_{k+1} = \Sigma\phi_i\gamma_i + \alpha\tau_{k+1}$; once again, this allows us to shorten our relation by one that holds in $E(Q)\otimes_N E(P)$, and by induction, the relation 1/ is a consequence of a relation in $E(Q)\otimes_n E(P)$.

We have shown that in all circumstances $\Sigma\beta_i\otimes\gamma_i + \delta = 0$, which is what we needed to prove our theorem.

To complete the proof of theorem 13.13, we need to show that this last result implies that inside $F' \underset{(M,x)}{\circ} M_n(E)$, $(M_n(E)F',1)$ is isomorphic to $(M_n(E)\otimes_F F',1\otimes 1)$.

We regard $M_2(F' \underset{(M,x)}{\circ} M_n(E))$ as the universal localisation of the matrix ring $T = \begin{pmatrix} F' & M \\ 0 & M_n(E) \end{pmatrix}$ at the standard u.t. rank function. Let $P = \begin{pmatrix} F & 0 \\ 0 & 0 \end{pmatrix}$ and let $Q = \begin{pmatrix} 0 & M \\ 0 & M_n(E) \end{pmatrix}$ and $\alpha:P \to Q$ be right multiplication by $\begin{pmatrix} 0 & x \\ 0 & 0 \end{pmatrix}$; then $End_R(Q) = M_n(E)$ and $End_R(P) = F'$; the isomorphism between $F' \underset{(M,x)}{\circ} M_n(E)$ and $End_{T_\rho}(T_\rho \otimes_T P)$ is induced by the map that sends F' to $End_R(P)$ and $M_n(E)$ to $\alpha End_R(Q)\alpha^{-1}$. Therefore, we need to calculate the pointed bimodule $(\alpha End_R(Q)\alpha^{-1}End_R(P),1)$ as $\alpha End_R(Q)\alpha^{-1}$, $End_R(P)$ bimodule; since α is an injective map, this is the same as calculating $(End_R(Q)\alpha^{-1}End_R(P),\alpha^{-1})$ as $End_R(Q)$, $End_R(P)$ bimodule. Since α is an atom, our last theorem shows that this isomorphic to $(End_R(Q)\otimes_N End_R(P),1\otimes 1)$ as pointed bimodule, where N is the normaliser of α; this is just the normaliser of $x \in M$ which we have shown is F. Therefore, $(M_n(E)F',1)$ is isomorphic to $(M_n(E)\otimes_F F',1\otimes 1)$.

This completes the proof of theorem 13.13 and, by symmetry, theorem 13.13' follows; we have already seen that theorem 13.12 follows from these.

An hereditary artinian ring of representation type $I_2(5)$

In (Dowbor, Ringel, Simson 79) they showed that hereditary artinian rings of finite representation type, by which we mean that there are only finitely many f.g. indecomposable modules, correspond to Coxeter diagrams in the same way that hereditary artinian algebras f.d. over a central subfield correspond to Dynkin diagrams. The Dynkin diagrams are all realised by suitable f.d. hereditary algebras, however, there were no known examples of hereditary artinian rings whose representation type corresponded to a Coxeter diagram that was not Dynkin since such example required an extension of skew fields $E \supset F$ where the left and right dimensions of E over F are finite and different; we shall construct an hereditary artinian ring of representation type corresponding to the Coxeter diagram $I_2(5)$. At present, there are no examples known corresponding to the other Coxeter diagrams apart from H_3 and H_4 which arise as a trivial consequence of the existence of suitable bimodules for $I_2(5)$.

A species is a directed graph whose vertices v are labelled by skew fields D_v and whose edges e are labelled by bimodules M_e where M_e is a D_{ie}, $D_{\tau e}$ bimodule for i_e the beginning and τ_e the end of the edge e. The tensor algebra of the species is a tensor algebra of the bimodule $\oplus_e M_e$ over the ring $X_v D_v$, where $D_v M_e = 0$ if $v \neq i_e$ and similarly $M_e D_v = 0$ if $v \neq \tau_e$. If R is an hereditary artinian ring of finite representation type, it is shown in the paper cited that R must be Morita equivalent to the tensor algebra of a species; so, the species may be associated to R in a natural way; next the bimodules that occur in the species of such an hereditary artinian ring of finite representation type may be assigned an integer in a natural way (corresponding to suitable linear data on the bimodule which we shall not explain in detail); replacing the species by the underlying undirected graph with the edges labelled by the integer associated to the given bimodule, we always obtain a graph whose connected components are Coxeter diagram. Hereditary artinian rings corresponding to the same Coxeter diagram have very similar module categories.

Next, we shall state precisely what conditions we need on a bimodule to construct an hereditary artinian ring of type $I_2(5)$. Let E and F be skew fields and let M be an E,F bimodule; we may form the F,E bimodule $M^R = \mathrm{Hom}_F(M,F)$; if $[M:F]_r = n$, then $[M^R:F] = n$ also. If the cardinals $[M:F]_r$, $[M^R:E]_r$, $[M^{RR}:F]_r$ ... are all finite, the bimodule is said to have finite right dualisation; the sequence of cardinals is always known as the right dimension sequence. We shall be interested in constructing a bimodule such that the sequence of cardinals begins 2,1,3,1,...; in (Dowbor, Ringel, Simson 79) it is shown that if such a bimodule exists the rest of the dimensions are determined; further, the ring $\begin{matrix} E & M \\ 0 & F \end{matrix}$ is an hereditary artinian ring of representation type $I_2(5)$. First, we turn this information on the bimodule into information on the skew fields E and F.

Let M be a G,H bimodule for skew fields G and H; if $[M:F]_r = n$, the action of G on M induces an embedding of G into $M_n(H)$; in turn we may recover M from an embedding of G into $M_n(H)$ by regarding the simple left $M_n(H)$ as G,H bimodule via the action of G induced by its embedding in $M_n(H)$; by the duality of rows and columns in a matrix ring, it is clear that $M^R = \mathrm{Hom}_H(M,H)$ is isomorphic as $H,M_n(H)$ bimodule and consequently as H,G bimodule to the simple right $M_n(H)$ module. Therefore, a bimodule sequence beginning $a,b,c \ldots$ for an H,G bimodule M corresponds to an embedding of G into $M_b(H)$ such that $[M_b(H):G]_l = ab$ whilst $[M_b(H):G]_r = cb$. In turn, if we have an E,F bimodule M which has a right bimodule sequence beginning $2,1,3,1,\ldots$ this corresponds to an embedding of F into E such that $[E:F]_l = 2$, $[E:F]_r = 3$, whilst for the embedding of E into $M_3(F)$ given by the left action of E on itself $[M_3(F):E]_l = 3$ and $[M_3(F):E]_r = 3$. It remains to construct such an extension of skew fields.

The construction

Let $E_O \supset F_O$ be an extension of skew fields such that $[E_O:F_O] = 2$, and $[E_O:F_O]_r = 3$; let $\{1,e\}$ be a left basis whilst $\{1 = e_1, e_2, e_3\}$ is a right basis. The left action of E on itself together with the given right basis induces an embedding of E_O into $M_3(F_O)$ where $x \in E_O$ is sent to the matrix (f_{ij}) such that $xe_j = \Sigma e_i f_{ij}$. Let $\{g_{ij}: i,j = 1 \text{ to } 3\}$ be matrix units corresponding to the basis chosen. Then $(M_3(F_O)g_{11}, g_{11})$ is isomorphic as pointed E_O, F_O bimodule via the given embedding of E_O into $M_3(F_O)$ to $(E_O, 1)$.

We shall construct skew fields

$$\begin{array}{ccc} \bar{E} & \supset & \bar{F} \\ \cup & & \cup \\ E_O & \supset & F_O \end{array}$$

such that $E_O \cap \bar{F} = F_O$, a left basis for $\bar{E}$ over F is $\{1,e\}$, a right basis for $\bar{E}$ over $\bar{F}$ is $\{e_1, e_2, e_3\}$; under the embedding of $\bar{E}$ into $M_3(\bar{F})$ given by the choice of basis (and therefore extending the embedding of E_O into $M_3(F_O)$) a left basis for $M_3(\bar{F})$ over $\bar{E}$ is $\{g_{ii}: i = 1 \text{ to } 3\}$ which is also a right basis.

At an odd stage of the construction, assume that we have an extension of skew fields

$$\begin{array}{ccc} E_{2m} & \supset & F_{2m} \\ \cup & & \cup \\ E_O & \supset & F_O \end{array}$$

such that a left basis for E_{2m} over F_{2m} is $\{1,e\}$ a right basis is $\{e_1, e_2, e_3\}$. Under the embedding of E_{2m} into $M_3(F_{2m})$ given by this choice of basis we have the extension of rings:

$$\begin{array}{ccc} M_3(F_{2m}) & \supset & E_{2m} \\ \cup & & \cup \\ M_3(F_O) & \supset & E_O \end{array}$$

such that a left basis for $M_3(F_{2m})$ over E_{2m} is $\{g_{ii}: i = 1 \text{ to } 3\}$. By theorem 13.12, there exist skew fields F_{2m+1}, E_{2m+1} and a diagram of rings:

$$\begin{array}{ccc} M_3(F_{2m+1}) & \supset & E_{2m+1} \\ \cup & & \cup \\ M_3(F_{2m}) & \supset & E_{2m} \end{array}$$

such that a left basis for $M_3(F_{2m+1})$ over E_{2m+1} is $\{g_{ii}: i = 1 \text{ to } 3\}$ which is also a right basis. We have an embedding of F_{2m+1} into E_{2m+1}, $f' \to f$ where $g_{11}f' = fg_{11}$ which extends the embedding of F_{2m} into E_{2m}. Under this embedding, $(E_{2m+1}, 1)$ as pointed E_{2m+1}, F_{2m+1} bimodule is isomorphic to $(M_3(F_{2m+1})g_{11}, g_{11})$. Since $\{e_1g_{11}, e_2g_{11}, e_3g_{11}\}$ is a right basis for $M_3(E_{2m})$ over F_{2m} and therefore a basis for $M_3(F_{2m+1})g_{11} \cong M_3(F_{2m})g_{11} \otimes_{F_{2m}} F_{2m+1}$, $\{e_1, e_2, e_3\}$ is also a right basis for E_{2m+1} over F_{2m+1}. Also, $E_{2m} \cap F_{2m+1} = F_{2m}$ since it is not E_{2m} and $[E_{2m}:F_{2m}]_r = 3$. It follows that $E_0 \cap F_{2m+1} = F_0$.

At an even stage of the construction, assume that we have an extension of skew fields:

$$\begin{array}{ccc} E_{2m-1} & \supset & F_{2m-1} \\ \cup & & \cup \\ E_0 & \supset & F_0 \end{array}$$

such that $E_0 \cap F_{2m-1} = F_0$, and a right basis for E_{2m-1} over F_{2m-1} is $\{e_1, e_2, e_3\}$. So, we have an embedding of E_{2m-1} into $M_3(F_{2m-1})$ extending our embedding of E_0 into $M_3(F_0)$. By theorem 13.12, there exist skew fields E_{2m} and F_{2m} and a diagram of skew fields:

$$\begin{array}{ccc} E_{2m} & \supset & F_{2m} \\ \cup & & \cup \\ E_{2m-1} & \supset & F_{2m-1} \end{array}$$

such that $E_{2m-1} \cap F_{2m} = F_{2m-1}$, a left basis for E_{2m} over F_{2m} is $\{1, e\}$ whilst $\{e_1, e_2, e_3\}$ is a right basis. Once more, we note that the embedding of E_{2m} into $M_3(F_{2m})$ induced by this right basis extends the embedding of E_{2m-1} into $M_3(F_{2m-1})$.

Consider the extension of skew fields, $\bar{E} \supset \bar{F}$, where $\bar{E} = \bigcup_i E_i$, and $\bar{F} = \bigcup_i F_i$. Any element of E lies in E_{2m-1} for some integer m and must be left dependent on $1, e$ and right dependent on e_1, e_2, e_3 over E_{2m}; if there were a dependence relation between 1 and e over $\bar{F}$, this would occur over some F_n which we know does not happen; so, $\{1, e\}$ is a left basis for $\bar{E}$ over $\bar{F}$ and similarly $\{e_1, e_2, e_3\}$ is a right basis. So, $[\bar{E}:\bar{F}] = 2$, and $[\bar{E}:\bar{F}]_r = 3$. This choice of basis induces an embedding of $\bar{E}$ into $M_3(\bar{F})$ which is just the union of the embeddings of E_n into $M_3(F_n)$ that we have considered at each stage of the construction. Any element of $M_3(\bar{F})$ lies in $M_3(F_{2m})$ for some integer m; therefore, it is both left and right dependent on $\{g_{ii} : i = 1 \text{ to } 3\}$ over E_{2m+1}. Clearly, there can be no dependence relation between the elements $\{g_{ii} : i = 1 \text{ to } 3\}$ over $\bar{E}$. So, $[M_3(\bar{F}):\bar{E}] = 3 = [M_3(\bar{F}):\bar{E}]_r$ as we wished.

REFERENCES

Amitsur, S.A. 1955. Generic splitting fields of central simple algebras. Annals of Maths. 62:8-43.

Bass, H. 1960. Finitistic dimension and a homological generalisation of semiprimary rings. Trans. Amer. Maths. Soc. 95:466-88.

Bass, H. 1968. Algebraic K-theory. Benjamin (New York).

Bass, H. 1979. Traces and Euler characteristics. Lond. Maths. Soc. Lecture notes 36:1-26.

Bass, H. & Murthy, M. 1967. Grothendieck groups and Picard groups of abelian group rings. Annals of Maths 86:16-73.

Bergman, G. 1967. Commuting elements in free algebras and related topics in ring theory. Ph.D. Thesis. Harvard.

Bergman, G. 1971. Hereditary and cohereditary projective modules. Park City symposium.

Bergman, G. 1974. Modules over coproducts of rings. Trans. Amer. Maths. Soc. 200:1-32.

Bergman, G. 1974a. Coproducts and some universal ring constructions. Trans. Amer. Maths. Soc. 200:33-88.

Bergman, G. 1976 Rational relations and rational identities in division rings, I, II. J. of Algebra 43:252-266, 267-297.

Bergman, G. & Dicks, W. 1975. On universal derivations. J. of Algebra 36:193-210.

Bergman, G. & Dicks, W. 1978. Universal derivations and universal ring constructions. Pacific J. of Maths 79:293-336.

Bergman, G. & Small, L. 1975. PI degrees and prime i-eals. J. of Algebra 33:433-462.

Björk, J.E. 1971. Conditions which imply that subrings of a semiprimary. J. of Algebra 19:384-395.

Burnside, W. 1911. The theory of groups of finite order. 2nd ed. Cambridge University Press.

Cameron, P. 1981. Finite permutation groups and finite simple groups. Bull. Lond. Maths. Soc. 13:1-22.

Cohn, P.M. 1969. Free associative algebras. Bull. Lond. Maths. Soc. 1:1-36.

Cohn, P.M. 1971. Free rings and their relations. Lond. Maths. Soc. Monographs 2. Academic Press (London, New York).

Cohn, P.M. 1974. Progress in free associative algebras. Israel J. of Maths. 19:109-149.

Cohn, P.M. 1977. Skew field constructions. London. Maths. Soc. Lecture Notes 27. Cambridge University Press.

Cohn, M.P. 1977a Centralisateurs dans les corps libre. Ecole de Printimps d'Informatique, Vieux-Boucains les Bains. (Landes).

Cohn, R.M. 1979. The affine scheme of a general ring. Springer Lecture Notes 753:197-211.

Cohn, P.M. 1982 The divisor group of a fir. Publ. Univ. Aut. Barcelona.

Cohn, P.M. 1984. Universal field of fractions of a semifir I, II, III (to appear in Proc. Lond. Maths. Soc.).

Cohn, P.M. & Dicks, W. 1976. Localisation in semifirs II. J. Lond. Maths. Soc. (2) 13:411-418.

Cohn, P.M. & Dicks, W. 1980. On central extension of skew fields. J. of Algebra 63:143-150.

Cohn, P.M. & Schofield, A.H. 1982. On the law of nullity. Proc. Camb. Phil. Soc. 91:357-374.

Dicks, W. 1977. Mayer-Vietoris presentations over colimits of rings. Proc. Lond. Maths. Soc. 34:557-576.

Dicks, W. 1980. Groups, trees and projective modules. Springer Lecture Notes 790.

Dicks, W. & Sontag, E. 1978. Sylvester domains. J. of Pure and Applied Algebra 13:243-275.

Dicks, W. & Lewin, J. 1982. A Jacobian conjecture for free associative algebras. Comm. in Algebra 10(72):1285-1306.

Dlab, V. & Ringel, C.M. 1976. Indecomposable representations of graphs and algebras. Mem. Amer. Maths. Soc. 173.

Dlab, V. & Ringel, C.M. 1984. A new class of hereditary noetherian bound domains. (to appear in J. of Algebra).

Dowbor, P., Ringel, C.M. & Simson, D. 1979. Hereditary artinian rings of finite representation type. Springer Lecture Notes 832:232-241.

Gersten, S. 1974. K-theory of free rings. Comm. in Algebra 1:39-64.

Gersten, S. 1975. The localisation theorem for projective modules. Comm. in Algebra 2(4):307-350.

Goodearl, K. 1979. Von Neumann regular rings. Pitman (London, San Francisco, Melbourne).

Isbell, J. 1969. Epimorphisms and dominions IV. J. Lond. Maths. Soc. (2)1: 265-273.

Lewin, J. 1969. Free modules over free algebras and free group algebras: the Schreier technique. Trans. Amer. Maths. Soc. 145:455-465.

Linnell, P. 1983. On accessibility of groups. J. of Pure and Applied Algebra 30:39-46.

Magnus, W., Karrass, A. & Solitar, D. 1966. Combinatorial group theory. Dover.

Malcolmson, P. 1980. Determining homomorphisms to skew fields. J. of Algebra 9:249-265.

Malcolmson, P. 1982. Construction of universal matrix localisation. Springer Lecture Notes 951:117-132.

Resco, R. 1980. Dimension theory for division rings. Israel J. of Maths. 35:215-221.

Resco, R., Small, L. & Wadsworth, A. 1979. Tensor products of division rings and finite generation of subfields. Proc. Amer. Maths. Soc. 77:7-10.

Revesz, G. 1984. On the abelianised multiplicative group of universal fields of fractions (to appear).

Ringel, C.M. 1978. Finite dimensional hereditary algebras of wild representation type. Maths. Zeit. 161:235-255.

Ringel, C.M. 1979. The simple artinian spectrum of a finite dimensional algebra. Dekker Lecture Notes 51:535-598.

Robson, C. 1968. Non-commutative Dedekind domains. J. of Algebra 9:249-264.

Roquette, P. 1962. On the Galois cohomology of the projective linear general linear group and its application to the construction of generic splitting fields of algebras. Maths. Ann. 150:411-439.

Rotman, J. 1979. An introduction to homological algebra. Academic Press (London, New York).

Silver, C. 1967. Non-commutative localisations and applications. J. of Algebra 7:44-76.

Tachikawa, M. 1973. Quasi-Frobenius rings and generalisations. Springer Lecture Notes Series 351.

Waterhouse, W. 1979. Introduction to affine group schemes (ch. 17 and 18). Graduate Texts in Maths. 66. Springer-Verlag, (New York).

INDEX